**Springer Theses**

Recognizing Outstanding Ph.D. Research

## Aims and Scope

The series "Springer Theses" brings together a selection of the very best Ph.D. theses from around the world and across the physical sciences. Nominated and endorsed by two recognized specialists, each published volume has been selected for its scientific excellence and the high impact of its contents for the pertinent field of research. For greater accessibility to non-specialists, the published versions include an extended introduction, as well as a foreword by the student's supervisor explaining the special relevance of the work for the field. As a whole, the series will provide a valuable resource both for newcomers to the research fields described, and for other scientists seeking detailed background information on special questions. Finally, it provides an accredited documentation of the valuable contributions made by today's younger generation of scientists.

## Theses are accepted into the series by invited nomination only and must fulfill all of the following criteria

- They must be written in good English.
- The topic should fall within the confines of Chemistry, Physics, Earth Sciences, Engineering and related interdisciplinary fields such as Materials, Nanoscience, Chemical Engineering, Complex Systems and Biophysics.
- The work reported in the thesis must represent a significant scientific advance.
- If the thesis includes previously published material, permission to reproduce this must be gained from the respective copyright holder.
- They must have been examined and passed during the 12 months prior to nomination.
- Each thesis should include a foreword by the supervisor outlining the significance of its content.
- The theses should have a clearly defined structure including an introduction accessible to scientists not expert in that particular field.

More information about this series at http://www.springer.com/series/8790

Sheon S.Y. Chua

# Quantum Enhancement of a 4 km Laser Interferometer Gravitational-Wave Detector

Doctoral Thesis accepted by
The Australian National University,
Canberra, Australia

Springer

*Author*
Dr. Sheon S.Y. Chua
Laboratoire Kastler Brossel
Université Pierre et Marie Curie
Paris
France

*Supervisor*
Prof. David McClelland
Department of Quantum Science, College of Physical and Mathematical Sciences
The Australian National University
Canberra
Australia

ISSN 2190-5053 ISSN 2190-5061 (electronic)
Springer Theses
ISBN 978-3-319-36702-6 ISBN 978-3-319-17686-4 (eBook)
DOI 10.1007/978-3-319-17686-4

Springer Cham Heidelberg New York Dordrecht London

Softcover reprint of the hardcover 1st edition 2015


Printed on acid-free paper

Springer International Publishing AG Switzerland is part of Springer Science+Business Media (www.springer.com)

*To my family and friends*

# Supervisor's Foreword

A global network of second-generation kilometre-scale interferometric detectors is expected to be operational in the next few years, with the aim of directly detecting gravitational-wave strain signals of the order of one part in $10^{23}$. These giant instruments, combining forefront technologies from many fields of engineering and physics, will be sensitivity-limited by fluctuations at the atomic and quantum level. One such limiting source is quantum fluctuations of laser light.

In the early 1980s, the injection of squeezed states of light was proposed by Caves as a way to reduce the impact of laser quantum fluctuations in gravitational-wave detectors. This theoretical work preceded the first experimental observation of squeezed light by 5 years. Since the first observation of squeezed light in the mid-1980s, the development of both squeezed light theory and squeezed light technology progressed steadily.

Even after 30 years of development, open questions remained. Two such questions had implications for the continuing viability of squeezed light as a technology for enhancing gravitational-wave detectors. The first was to be able to generate large magnitudes of squeezed light across the measurement Fourier band of gravitational-wave detectors—10 Hz–10 kHz. The second was to verify the capability of using squeezed light at the measurement sensitivities and realistic operating conditions of gravitational-wave detectors. A squeezed light test on a full-scale interferometer was needed.

Early in his work, Sheon Chua diagnosed the performance of a squeezed light generator to determine why (1) we were limited to 8 dB of squeezing and (2) we could only observe squeezing down to 100 Hz. He, along with fellow Ph.D. student Michael Stefszky, implemented a new design and through very careful work was able to demonstrate a squeezed light source achieving more than 11 dB of measured squeezing (17 dB generated) down to 10 Hz. This squeezed light source has all the characteristics required of the source needed for enhancing the performance of long baseline gravitational wave detectors.

Injecting a squeezer into a 4 km long interferometer is not a single researcher benchtop experiment. In 2011, Sheon joined the Mavalvala's group at Massachusetts

Institute of Technology in a project to integrate a squeezer built at the Australian National University into one of the 4 km long LIGO detectors as a proof of principle of the technique. He was one of the key scientists in this work along with Drs Barsotti and Sigg and then MIT Ph.D. student Sheila Dwyer with extensive support for LIGO Hanford staff. The experiment was a resounding success, delivering what was then the most sensitive interferometer ever constructed. The importance of this work cannot be underestimated. With the squeezer Sheon was instrumental in building and the implementation demonstrated on LIGO, in which he played a key role, advanced detectors will either reach their design sensitivity at reduced (by up to a factor of four) laser power or have their science reach doubled!

In his thesis, Sheon presents a clear overview of squeezing including an extensive literature review, followed by a comprehensive description and critical analysis of what it takes to build and use low frequency squeezers. He analyses performance limits and ends with a concise summary of what now needs to be done. His thesis will serve as an ideal reference for future students in the field.

This thesis has been awarded the Gravitational Wave International Committee (GWIC) Thesis Prize for 2013.

Canberra, Australia
November 2014

David McClelland
Centre for Gravitational Physics,
The Australian National University

# Preface

The ability to directly detect gravitational waves will open a completely new branch of astronomy to view the Universe, one that is inaccessible to electromagnetic-based astronomy. First-generation ground-based interferometric gravitational-wave detectors have achieved strain sensitivities of order $10^{-21}$, at 100 Hz detection frequency. A new generation of detectors is under construction, designed to improve on the sensitivity of the first-generation detectors by a factor of 10.

The quantum nature of light will broadly limit the sensitivity of these new instruments. This quantum noise will originate from the quantum vacuum fluctuations that enter the unused port of the interferometer. One of the most promising options for reducing the quantum noise impact and further increasing the sensitivity is applying quantum squeezed vacuum states. These squeezed states have lower noise in one quadrature than the vacuum state. By replacing the quantum vacuum fluctuations entering the interferometer with squeezed vacuum states, the quantum noise impact is reduced.

This thesis is an account of research in the field of squeezed states for enhancing interferometric gravitational-wave detectors, undertaken between February 2008 and March 2013, administered by the Department of Quantum Science, Research School of Physics and Engineering, The Australian National University, Canberra, Australia. This Springer Theses Edition 2015 is a slightly revised version of the original thesis of April 2013.

Following the introduction and thesis overview of Chap. 1, the thesis is divided into three parts. Part I provides the background theory for the research presented, with material on the characteristics of gravitational waves and their detection (Chap. 2), quantum optics (Chap. 3), quantum noise limits of interferometric gravitational-wave detectors (Chap. 4) and squeezed-light sources (Chap. 5).

In Part II, Chap. 6 presents a doubly-resonant, travelling-wave bow-tie cavity squeezed light source and development work leading to the first measurement of $11.6 \pm 0.4$ dB of quantum noise suppression from 200 Hz and above. The properties affecting squeezing magnitude and low-frequency squeezing measurement are discussed. In addition, a modified squeezing-ellipse-phase control technique for squeezed vacuum states is presented. This squeezing cavity design also has benefits with intrinsic isolation to backscattered light. Chapter 7 presents the measurement of the immunity of the travelling-wave Optical Parametric Oscillator to backscattered light. The immunity value, measured using the balanced homodyne detector, is compared to a squeezing-noise coupling spectrum measurement.

Part III presents the methodologies and results from the H1 Squeezed Light Injection Experiment, performed at the Laser Interferometer Gravitational-Wave Observatory (LIGO). Chapter 8 introduces the project background, interferometer hardware modifications made for the squeezing injection, and squeezed light source, based on the Optical Parametric Oscillator from Part II of this thesis. Chapter 9 presents the main results of the LIGO H1 Squeezed Light Experiment, particularly the 2.15 ± 0.05 dB squeezing enhancement of the 4 km Enhanced LIGO interferometer above 250 Hz. An unknown area was whether the addition of a squeezer would introduce noise couplings that degrade the crucial low frequency sensitivity. Chapter 10 presents the measurements made to investigate the impact of backscattered light from an installed squeezer on LIGO. Inferred levels of backscatter noise on the interferometer readout, made from vibration excitations and optical path length modulations, are presented.

The knowledge and processes gained, from both the squeezed light source development work and the LIGO Squeezed Light Injection Experiment, will inform the design, planning and implementation of squeezed states in future gravitational-wave detectors. Chapter 11 concludes the thesis, with a summary of the work presented and the discussion of extensions to this research.

Canberra, Australia
December 2014

Sheon S.Y. Chua

# Acknowledgments

There are many talented and wonderful people that I have worked with over my Ph.D., whom I wish to sincerely thank.

First and foremost, I thank my Chair Supervisor, Professor David McClelland. Your knowledge and wisdom, spanning the finest detail to the 'global' picture, is second to none. Your patience and your willingness to make time for discussions was critical to the success of my Ph.D. As group leader of the ANU Centre for Gravitational Physics, you have created a research atmosphere that is productive and friendly, well suited for scientific achievement. To my other supervisors, Profs. Daniel Shaddock and Ping Koy Lam. To Daniel, thank you for your different viewpoints in approaching problems, and the 'crazy ideas' that inevitably turn out to be useful and correct. To Ping Koy, thank you for your vast knowledge in quantum optics, preventing many 'reinvention of the wheel' moments.

To my lab supervisors, Drs. Bram Slagmolen and Ben Buchler. Thank you for your perpetual willingness to drop everything you were doing to help me in the lab, as well as answer the full range of specialised and mundane questions.

To my lab partners, Michael Stefszky and Conor Mow-Lowry. To Michael, thank you for providing a whole different perspective to solving problems, your work ethic and your positive attitude. It is no surprise that working together with you meant that the squeezed-light source appeared to 'spontaneously' optimise and operate, with our near-instinctive process, multi-tasking and efficiency. To Conor, thank you for your continuous stream of solutions to all manner of technical issues that arose.

To my fellow 'Graviteers': James Dickson, Jong Chow, Timothy Lam, Roland Fleddermann, John Miller, Ra Inta, Alberto Stochino, Robert Ward, Kerrie Cook, Thanh Nguyen, Andrew Sutton, Danielle Wuchenich, Dave Bowman, Andrew Wade, Silvie Ngo and Lyle Roberts. Thank you for the spontaneous discussions of physics, brainstorming sessions to solve experiment issues, the friendly smiles, and also my gratitude to the great number of you that have provided feedback for various parts of this thesis. I am privileged to be a Graviteer alongside you.

To my supervisors at LIGO Hanford Observatory, Drs. Lisa Barsotti, Daniel Sigg and Keita Kawabe. To Lisa, thank you for your endless expertise, patience and dedication to achieving scientific outcomes. Without your leadership, the LIGO Squeezed Light Injection Experiment would not have been a total success. To Daniel, thank you for your depth of knowledge that expanded my knowledge of physics and electronics, and your near-instantaneous solutions to problems. Thank you also for the 'Dessert of the Century' that you shared with those like myself—researchers who were far from home on Thanksgiving. To Keita, thank you for imparting some of your vast insights into interferometer operation, and ensuring installations within the interferometer were smooth and efficient.

To Sheila Dwyer from MIT, whom I spent alongside the vast majority of the late nights, early mornings, all-nighters and weekends operating the H1 squeezed-light source and Enhanced LIGO interferometer. Thank you for your seemingly endless energy, attention to detail in your work and your positive attitude. It is my privilege to call you my colleague and peer.

To my fellow researchers of the LIGO Squeezed Light Injection team—Nergis Mavalvala, Michael Landry, Robert Schofield, Matthew Evans, Cheryl Vorvick, Richard Gustafson, Grant Meadors, Maxim Factourovich, Alexander Khalaidovski and Nick Smith-Lefebvre. Thank you for your contributions essential to the success of the project. In particular, thank you to Nergis Mavalvala at MIT and Michael Landry at Hanford, whom without making available the resources and time allocation, would have made the squeezing experiment impossible to complete.

To the wonderful staff at the LIGO Hanford Observatory and the LIGO Scientific Collaboration, particularly to Fred Raab, Dale Ingram, Greg Grabeel, Filiberto Clara, Jonathan Hanks, Valera Frolov, Sam Waldman, Gerardo Moreno, Bubba Gateley, Michael Rodruck, Thomas Vo, Carolyn Peterson, Cris Murphy, Karen Boland—thank you for welcoming me and making my stay so enjoyable, productive and memorable. With such a talented and friendly crew, scientific success is assured.

Lastly, but certainly not least, thank you to all my family and friends. To the "old boys": Sanjit, John, Mark, Dave, Dan, Paul, Luke, Thomas and Anthony—thank you for keeping life real. To my Dad, my Mum and my brother Andrew, thank you for all your love and support on this journey.

# Contents

# Chapter 1
# Introduction

**Abstract** The ability to directly detect gravitational waves will open a completely new branch of astronomy to view the Universe, one that is inaccessible to electromagnetic-based astronomy (Thorne, Gravitational Waves, 1995, [1]). The development of that capability has many technical challenges. Finding solutions to these challenges have called upon the research development of many different fields of science and engineering. The first devices for the direct detection of gravitational waves, known collectively as the *resonant mass detectors*, were being proposed and constructed from 1960, starting with the work of Weber (Phys Rev 117:306, 1960, [2]). Contemporary to this, the field of quantum optics was beginning, with the formalism work of Glauber (Phys Rev 131:2766–2788, 1963, [3]). The cross-over between these fields began from the foundation work in quantum-noise-limited measurement that was being driven by the gravitational-wave detection field (Sov Phys JETP 46:705–706, 1977; JETP Lett 27:276–280, 1978, [4, 5]), as resonant mass detectors developed with increasing sensitivity that approached quantum limits. The use of laser interferometers for gravitational-wave detection was first raised in 1962 by Gertsenshtein and Pustovoit (JETP 43:605–607, 1962, [6]). The first in-depth scientific study into the construction of these *interferometric gravitational-wave detectors* came in 1972 by Weiss (Q Prog Rep Res Lab Electron 105, 1972, [7]). However, it was soon realised that these instruments would be sensitivity-limited by the quantum nature of the laser light itself. This lead to the theoretical study of the use of quantum *squeezed states for enhancing interferometric gravitational-wave detector sensitivity* (Other quantum measurement schemes being developed include variational interferometer readout (Kimble et al., Phys Rev D 65:022002, 2002; Corbitt and Mavalvala, J Opt B Quantum Semiclass Opt 6:S675, 2004, [8, 9]), 'speed-meter' designs (Purdue and Chen, Phys Rev D 66:122004, 2002; de Vine et al., Phys Lett A 316:17–23, 2003; Chen, Phys Rev D 67:122004, 2003; Chen et al., General Relativ Gravity 43:671–694, 2011; Wade et al., Phys Rev D 86:062001, 2012, [10–14]), which are acknowledge but not presented.). It is this quantum-optic enhancement scheme that is the focus of this thesis.



S.S.Y. Chua, *Quantum Enhancement of a 4 km Laser Interferometer Gravitational-Wave Detector*, Springer Theses,
DOI 10.1007/978-3-319-17686-4_1

## 1.1 Squeezed States for Gravitational-Wave Detection

Squeezed states for the enhancement of interferometric gravitational-wave detectors was first proposed by Caves in 1981 [15]. This proposal was further built on by Unruh [16] and later Jaekel and Reynaud [17] by showing that a squeezed state (with appropriate squeezing angle) can reduce the noise below the lower-limit imposed by the quantum nature of light. The theory of squeezed states will be covered in Chap. 3, and its application in interferometric gravitational-wave detectors in Chap. 4.

From the first observation of squeezing by Slusher et al. [18], progress in both generating larger squeezing magnitudes, and measuring squeezed states at lower Fourier frequencies for gravitational-wave detector enhancement has been steady. This development is charted in Fig. 1.1.

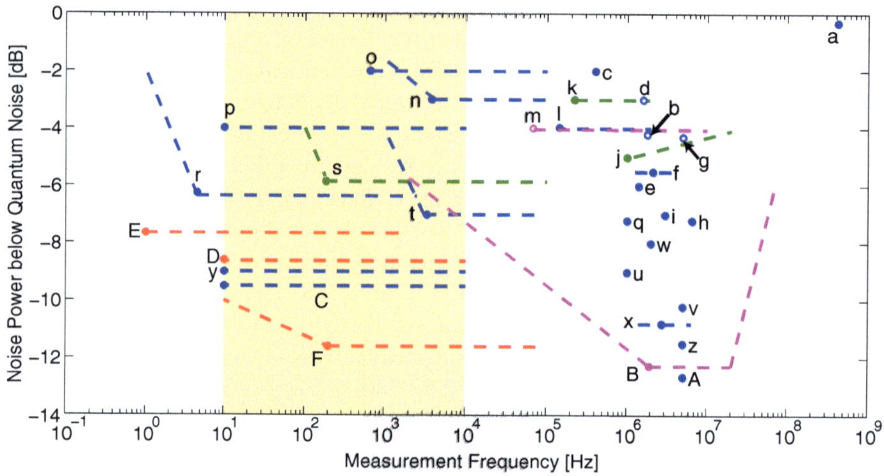


**Fig. 1.1** Plot of previous measured squeezing magnitudes and measurement Fourier frequencies. The audio gravitational-wave detection band (10 Hz–10 kHz) is shown by the shaded area. Differing colours are for clarity only. *a* First squeezing measurement—Slusher et al. [18], *b* Wu et al. [19], *c* Grangier et al. [20], *d* Xiao et al. [21], *e* Polzik et al. [22], *f* Breitenbach et al. [23], *g* Schneider et al. [24], *h* Schneider et al. [25], *i* Lam et al. [26], *j* Buchler et al. [27], *k* Bowen et al. [28], *l* Schnabel et al. [29], *m* Laurat et al. [30], *n* first squeezing in audio GW band—McKenzie et al. [31], *o* McKenzie et al. [32], *p* Vahlbruch et al. [33], *q* Suzuki et al. [34], *r* first squeezing across the audio GW band—Vahlbruch et al. [35], *s* McKenzie [36], *t* Goda et al. [37], *u* Takeno et al. [38], *v* Vahlbruch et al. [39], *w* Yonezawa et al. [40], *x* Vahlbruch [41], *y* Vahlbruch et al. [42], *z* Mehmet et al. [43], *A* 12.6 dB squeezing at 5.5 MHz—Eberle et al. [44], *B* 12.3 dB at 1550 nm—Mehmet et al. [45], *C* Khalaidovski et al. [46], *D* Chua et al. [47], *E* this thesis, *F* 11.6 dB at 200 Hz, >10 dB across the audio GW band—Stefszky et al. [48]

### 1.1.1 Squeezing Magnitude

Measured squeezing magnitude has steadily increased over time from refinements in experimental technique and optical components. The largest magnitudes in squeezing yet achieved have been with the squeezing apparatus that use nonlinear optical media. A squeezing magnitude of 12.6 dB has been observed by the group at the Albert-Einstein Institute (AEI) in Hannover, Germany [44].

### 1.1.2 Audio-Detection Band Fourier Frequencies

Squeezing measurement in the gravitational-wave detection band is a key requirement for squeezing to be useful for gravitational-wave sensitivity enhancement. For a long period after first measurement by Slusher et al., detection frequencies remained above the audio gravitational-wave detection band. In 2004, McKenzie et al. made the first measurement of squeezing within the detection band by connecting the relationship between the presence of bright field at the squeezing frequency and technical noise coupling [31]. Shortly after, first measurement across the entire detection band was achieved by Vahlbruch et al. [33]. Subsequent studies of the issues surrounding squeezed state detection in the audio band were made and reported on [35, 49], leading to the comprehensive study of Stefszky et al. [48], joint work between the Australian National University (ANU) and the Albert-Einstein Institute (AEI) in Hannover, Germany. This work also contains the first measurement of greater than 10 dB squeezing across the audio detection band.

### 1.1.3 Proof-of-Principle Experiments and Application in Gravitational-Wave-Detection Interferometers

Proof-of-principal experiments for squeezed light injection into benchtop and prototype interferometers in gravitational-wave-detector configuration have been made.

- Bench-top Interferometers—The first experiment with squeezing-enhanced signal detection from a simple Michelson interferometer with power-recycling was reported by McKenzie et al. [50], where the demonstrated enhancement was ~3 dB with an injected 5.46 MHz signal into the Michelson. The first demonstration of squeezing-enhanced detection with a dual-recycled Michelson interferometer was performed by Vahlbruch et al. [51], showing enhancement of up to 2.8 dB between 5 and 15 MHz.
- Prototype Interferometer—A milestone in the implementation of squeezing injection into gravitational-wave detectors was the successful test on the Caltech 40 m prototype interferometer in 2008 [52]. This interferometer is a miniaturised copy of a gravitational-wave detector, with suspended optics and recycling mirrors. The prototype injection saw an improvement of 3 dB at the kHz frequencies.

Key milestones in the application of squeezed states in large-scale interferometric gravitational-wave detectors have been reached:

- Implementation into GEO600—The GEO600 interferometer, located near Hannover, Germany was the first to implement squeezed state injection as part of baseline design and operation. An improvement of 3.5 dB above 700 Hz was reported in 2011 [53].
- The LIGO H1 Squeezed Light Injection Experiment, undertaken at the LIGO Hanford Observatory in Washington State, USA, confirmed the sensitivity-enhancement capability of squeezing around the 100 Hz audio detection frequency.

## 1.2 Thesis Overview

The development of the doubly-resonant, travelling-wave squeezed light source resulting in greater than 10 dB squeezing magnitude measured across the audio detection band, and the results of the LIGO H1 Squeezing Light Injection Experiment, form the major sections of research presented in this thesis. The structure of the thesis comprises of three major parts.

Part 1 provides the background theory for the research presented. This part calls upon published texts and reference material. This part is intended for completeness. For the reader familiar with these topics, this first part can be bypassed.

- Chapter 2 provides an introduction to gravitational waves and their detection. An overview of gravitational-wave sources, current and future gravitational-wave detectors, and an overview into sensitivity-limiting noise sources is presented. Pitkin et al. [54] is recommended for further detail.
- Chapter 3 presents the basics of quantum states of light and their pictorial representations, along with the quantum descriptions of optical components such as mirrors, optical cavities and photodetection. This chapter is based on the comprehensive text of Walls and Milburn [55].
- Chapter 4 introduces the calculations for obtaining the sensitivity-limiting quantum noise curves for interferometric gravitational-wave detectors. Three detector configurations are introduced, as well as the effect of applied squeezing for each configuration. The calculations presented are heavily based on the works of Kimble et al. [8], and Buonanno and Chen [56, 57].
- Chapter 5 gives an overview of a 'generic' squeezed-light source used for interferometric gravitational-wave detector enhancement. The various key parts are canvassed, including the laser inputs, nonlinear optics, measurement, stabilisation and control stages. The reference material for further detail are the works of Buchler [27], McKenzie [36] and Vahlbruch [41].

In Part 2, the development work and results from the doubly-resonant, travelling-wave squeezed light source are presented.

- Chapter 6 details the doubly-resonant, travelling-wave squeezed light source. The Optical Parametric Oscillator and the modified squeezing-angle control scheme are presented. The properties that affect squeezing magnitude and low-frequency measurement are then presented. The progression of upgrades to the squeezed light source, reflecting the progression of knowledge gained into these properties, are outlined. Results of up to 11.6 dB squeezing in the audio gravitational-wave detection band (Curve (E) of Fig. 1.1), and 5900 s of continuous 7.5 dB squeezing, down to 1 Hz (Curve (D) of Fig. 1.1) are reported.
- Chapter 7 presents the measurement of the immunity of the travelling-wave Optical Parametric Oscillator to backscattered light. The immunity value, measured with using the balanced homodyne detector, is compared to a squeezing-noise coupling spectrum measurement.

Part 3 presents the methodologies and results from the LIGO H1 Squeezed Light Injection Experiment.

- Chapter 8 introduces the LIGO H1 Squeezed Light Injection Experiment. The project background, the Enhanced LIGO interferometer and the hardware modifications made for the squeezing injection are presented. The LIGO H1 squeezed light source is highlighted, based on the doubly-resonant, travelling-wave Optical Parametric Oscillator.
- Chapter 9 presents the main results of the LIGO H1 Squeezed Light Experiment, particularly the sensitivity enhancement of the 4 km Enhanced LIGO interferometer using squeezed states. This includes the first measurement of squeezing-enhancement below 300 Hz detection frequency in an interferometric gravitational-wave detector, and the best sensitivity to gravitational waves above 250 Hz.
- Chapter 10 presents the measurements made to investigate the impact of backscattered light from an installed squeezer on LIGO. Inferred levels of backscatter noise on the interferometer readout, made from vibration excitations and optical path length modulations, are presented.

Chapter 11 concludes the thesis, with a summary of the work presented and the discussion of extensions to this research.

## 1.3 Statement of Contribution

### *Research Work at the Australian National University*

The work at the Australian National University was jointly completed with M. Stefszky and C. Mow-Lowry, under the supervision of D. McClelland, P.K. Lam, B. Buchler and D. Shaddock. This work was published in [47, 48].

I led the Optical Parametric Oscillator tolerance to backscatter, medium-term length squeezing, and squeezing ellipse phase noise measurement work. Squeezing spectrum measurements were made jointly with M. Stefszky. The work on characterising low-frequency noise sources in squeezing detection was led by M. Stefszky and C. Mow-Lowry, with my input. Mitigation of low-frequency noise on the squeezed light source was jointly implemented by M. Stefszky, C. Mow-Lowry and me. The modified coherent sideband locking scheme was devised by C. Mow-Lowry and K. McKenzie, with experimental implementation led by me.

The homodyne detector electronics for the later squeezing experiments was designed by H. Vahlbruch and R. Schnabel, from the Albert Einstein Institute, Hannover Germany.

### *Research Work at the LIGO Hanford Observatory*

The LIGO H1 Squeezed Light Injection Experiment was a collaboration of many members of the LIGO Scientific Collaboration. The lead scientists over the Experiment window were myself, S. Dwyer, L. Barsotti and D. Sigg, with invaluable contributions from M. Factourovich, G. Meadors, M. Stefszky, A. Khalaidovski, and C. Mow-Lowry.

The LIGO squeezing Optical Parametric Oscillator was constructed and tested by myself, M. Stefszky, C. Mow-Lowry, S. Dwyer and A. Khalaidovski, under the supervision of L. Barsotti, D. Sigg, K. Kawabe, R. Schnabel, D. McClelland and N. Mavalvala. Integration with the LIGO interferometer was overseen by Kawabe, with invaluable support from Landry and the LIGO Hanford Observatory staff. N. Smith-Lefebvre, M. Evans, R. Schofield, C. Vorvick, and R. Gustafson kept the LIGO interferometer running at peak sensitivity. The running of the squeezer was the responsibility of myself, Dwyer and Barsotti.

The backscatter impact on interferometer sensitivity was led by myself and R. Schofield, with S. Dwyer, L. Barsotti and D. Sigg. Interferometer sensitivity traces with squeezing were jointly completed by myself, S. Dwyer, L. Barsotti, D. Sigg, M. Evans and K. Kawabe, with input from G. Meadors, M. Factourovich and N. Smith-Lefebvre. Injection characterisation work was completed by S. Dwyer, myself, L. Barsotti, M. Evans, D. Sigg and K. Kawabe. Squeezing phase noise work on LIGO was led by Dwyer, with input from myself, Barsotti, and Sigg.

## 1.4 Publications

- Stefszky, M.S, Mow-Lowry, C.M., **Chua, S.S.Y.**, Shaddock, D.A., Buchler, B.C., Vahlbruch, H., Khalaidovski, A., Schnabel, R., Lam, P.K., and McClelland, D.E., *Balanced Homodyne Detection of Optical Quantum States at Audio-Band Frequencies and Below*, Class. Quant. Grav., vol. 29, no. 14, 145015 (2012)

- **Chua, S.S.Y.**, Stefszky, M.S., Mow-Lowry, C.M., Buchler, B.C., Dwyer, S., Shaddock, D.A., Lam, P.K. and McClelland, D.E., *Backscatter tolerant squeezed light source for advanced gravitational-wave detectors*, Opt. Lett., vol. 36, no. 23, pp. 4680–4682 (2011)
- **Chua, S.**, Stefszky, M., Mow-Lowry, C., Buchler, B.C., McKenzie, K., Shaddock, D.A., Lam, P.K. and McClelland, D.E., *Quantum squeezing in advanced gravitational wave detectors*, Int. J. Mod. Phys. D, vol. 20, no. 10, pp. 2043–2049 (2011)
- Stefszky, M., Mow-Lowry, C.M., McKenzie, K., **Chua, S.**, Buchler, B.C., Symul, T., McClelland, D.E. and Lam, P.K, *An investigation of doubly-resonant optical parametric oscillators and nonlinear crystals for squeezing*, J. Phys. B Atomic, Molecular and Optical Physics, vol. 44, no. 1., 015502 (2011)
- De Vine, G., Rabeling, D.S., Slagmolen, B.J.J., Lam, T.T.-Y., **Chua, S.**, Wuchenich, D.M., McClelland, D.E. and Shaddock, D.A., *Picometer level displacement metrology with digitally enhanced heterodyne interferometry*, Opt. Exp., vol. 17, no. 2, pp. 828–837 (2009)

- LIGO Scientific Collaboration—Author acknowledged as a lead scientist, *Enhanced sensitivity of the LIGO gravitational wave detector by using squeezed states of light*, Nat. Phot., vol. 7, pp. 613–619 (2013)
- **Chua, S.S.Y.**, Dwyer, S., Barsotti, L., Sigg, D., Schofield, R.M.S., Frolov, V.V., Kawabe, K., Evans, M., Meadors, G.D., Factourovich, M., Gustafson, R., Smith-Lefebvre, N., Vorvick, C., Landry, M., Khalaidovski, A., Stefszky, M.S., Mow-Lowry, C.M., Buchler, B.C., Shaddock, D.A., Lam, P.K., Schnabel, R., Mavalvala, N., and McClelland, D.M.,*Impact of backscattered-light in a squeezing-enhanced interferometric gravitational-wave detector*, Class. Quant. Grav., vol. 31, no. 14, 035017 (2014)
- Dwyer, S., Barsotti, L., **Chua, S.S.Y.**, Evans, M., Factourovich, M., Gustafson, D., Isogai, T., Kawabe, K., Khalaidovski, A., Lam, P.K., Landry, M., Mavalvala, N., McClelland, D.M., Meadors, G.D., Mow-Lowry, C.M., Schnabel, R., Smith-Lefebvre, N., Stefszky, M.S., Vorvick, C., and Sigg, D., *Squeezed quadrature fluctuations in a gravitational wave detector using squeezed light*, Opt. Exp., vol. 21, no. 16, 19047–19060 (2013)

## References


1. K.S. Thorne, Gravitational Waves (1995), arXiv:gr-qc/9506086v1
2. J. Weber, Detection and generation of gravitational waves. Phys. Rev. **117**, 306 (1960)
3. R.J. Glauber, Coherent and incoherent states of the radiation field. Phys. Rev. **131**, 2766–2788 (1963)
4. V.B. Braginsky, Y.I. Vorontsov, F.Y. Khalili, Quantum singularities of a ponderomotive meter of electromagnetic energy. Sov. Phys. JETP **46**, 705–706 (1977)
5. V.B. Braginsky, Y.I. Vorontsov, F.Y. Khalili, Optimal quantum measurements in detectors of gravitation radiation. JETP Lett. **27**, 276–280 (1978)

6. M.E. Gertsenshtein, V.I. Pustovoit, On the detection of low frequency gravitational waves. JETP **43**, 605–607 (1962)
7. R. Weiss, Electromagnetically coupled broad-band gravitational wave antenna. Q. Prog. Rep. Res. Lab. Electron. **105**(54) (1972)
8. H.J. Kimble, Y. Levin, A.B. Matsko, K.S. Thorne, S.P. Vyatchanin, Conversion of conventional gravitational-wave interferometers into quantum nondemolition interferometers by modifying their input and/or output optics. Phys. Rev. D **65**, 022002 (2002)
9. T. Corbitt, N. Mavalvala, Review: quantum noise in gravitational-wave interferometers. J. Opt. B Quantum Semiclass. Opt. **6**(8), S675 (2004)
10. P. Purdue, Y. Chen, Practical speed meter designs for quantum nondemolition gravitational-wave interferometers. Phys. Rev. D **66**, 122004 (2002)
11. G. de Vine, M. Gray, D.E. McClelland, Y. Chen, S. Whitcomb, Measurement of the frequency response of a bench-top quantum speed meter interferometer. Phys. Lett. A **316**(1–2), 17–23 (2003)
12. Y. Chen, Sagnac interferometer as a speed-meter-type, quantum-nondemolition gravitational-wave detector. Phys. Rev. D **67**, 122004 (2003)
13. Y. Chen, S.L. Danilishin, F.Y. Khalili, H. Müller-Ebhardt, QND measurements for future gravitational-wave detectors. General Relativ. Gravity **43**, 671–694 (2011)
14. A.R. Wade, K. McKenzie, Y. Chen, D.A. Shaddock, J.H. Chow, D.E. McClelland, Polarization speed meter for gravitational-wave detection. Phys. Rev. D **86**, 062001 (2012)
15. C.M. Caves, Quantum-mechanical noise in an interferometer. Phys. Rev. D **23**(8) (1981)
16. W.G. Unruh, *Quantum Optics, Experimental Gravitation and Measurement Theory*, ed. by P. Meystre, M.O. Scully (Plenum, New York, 1983), p. 64
17. M.T. Jaekel, S. Reynaud, Quantum limits in interferometric measurements. EPL (Europhys. Lett.) **13**(4), 301 (1990)
18. R. Slusher, L. Hollberg, B. Yurke, J. Mertz, J. Valley, Observation of squeezed states generated by four-wave mixing in an optical cavity. Phys. Rev. Lett. **55**, 22 (1985)
19. L.-A. Wu, H. Kimble, J. Hall, H. Wu, Generation of squeezed states by parametric down conversion. Phys. Rev. Lett. **57**, 20 (1986)
20. P. Grangier, R. Slusher, B. Yurke, A. LaPorta, Squeezed-light-enhanced polarization interferometer. Phys. Rev. Lett. **59**, 19 (1987)
21. M. Xiao, L.-A. Wu, H. Kimble, Precision measurement beyond the shot-noise limit. Phys. Rev. Lett. **59**, 3 (1987)
22. E. Polzik, J. Carri, H. Kimble, Atomic spectroscopy with squeezed light for sensitivity beyond the vacuum-state limit. Appl. Phys. B **55**, 279–290 (1992)
23. G. Breitenbach, T. Müller, S. Pereira, J.-Ph. Poizat, S. Schiller, J. Mlynek, Squeezed vacuum from a monolithic optical parametric oscillator. J. Opt. Soc. Am. B **12**, 11 (1995)
24. K. Schneider, R. Bruckmeier, H. Hansen, S. Schiller, J. Mlynek, Bright squeezed-light generation by a continuous-wave semimonolithic parametric amplifier. Opt. Lett. **21**, 17 (1996)
25. K. Schneider, M. Lang, J. Mlynek, S. Schiller, Generation of strongly squeezed continuous-wave light at 1064 nm. Opt. Express **21**, 3 (1998)
26. P.K. Lam, T. Ralph, D. McClelland, B. Buchler, H.-A. Bachor, J. Gao, Optimization and transfer of vacuum squeezing from an optical parametric oscillator. J. Opt. B. Quantum Semiclass Opt. **1**, 469–474 (1999)
27. B.C. Buchler, Electro-optic control of quantum measurements. Ph.D. thesis, Physics Department, Australian National University, Canberra, Australia, 2008
28. W. Bowen, R. Schnabel, N. Treps, H.-A. Bachor, P.K. Lam, Recovery of continuous wave squeezing at low frequencies. J. Opt. B Quantum Semiclass Opt. **4**, 421–424 (2002)
29. R. Schnabel, H. Vahlbruch, A. Franzen, N. Grosse, S. Chelkowski, H.-A. Bachor, W.P. Bowen, P.K. Lam, K. Danzmann, Squeezed light at sideband frequencies below 100 kHz from a single OPA. Opt. Commun. **240**, 185–190 (2004)
30. J. Laurat, T. Coudreau, G. Keller, N. Treps, C. Fabre, Compact source of Einstein-Podolsky-Rosen entanglement and squeezing at very low noise frequencies. Phys. Rev. A **70**, 042315 (2004)

31. K. McKenzie, N. Grosse, W. Bowen, S. Whitcomb, M. Gray, D. McClelland, P.K. Lam, Squeezing in the audio gravitational-wave detection band. Phys. Rev. Lett. **93**(16), 161105 (2004)
32. K. McKenzie, M. Gray, S. Goßler, P.K. Lam, D. McClelland, Squeezed state generation for interferometric gravitational-wave detection. Class. Quantum Gravity **23**, S245S250 (2006)
33. H. Vahlbruch, S. Chelkowski, B. Hage, A. Franzen, K. Danzmann, R. Schnabel, Coherent control of vacuum squeezing in the gravitational-wave detection band. Phys. Rev. Lett. **97**, 011101 (2006)
34. S. Suzuki, H. Yonezawa, F. Kannaru, M. Sasaki, A. Furusawa, 7 dB quadrature squeezing at 860 nm with periodically poled $KTiOPO_4$. Appl. Phys. Lett. **89**, 061116 (2006)
35. H. Vahlbruch, S. Chelkowski, K. Danzmann, R. Schnabel, Quantum engineering of squeezed states for quantum communication and metrology. New J. Phys. **9**, 371 (2007)
36. K. McKenzie, Squeezing in the audio gravitational wave detection band. Ph.D. thesis, Physics Department, Australian National University, Canberra, Australia, 2008
37. K. Goda, E. Mikhailov, O. Miyakawa, S. Saraf, S. Vass, A. Weinstein, N. Mavalvala7, Generation of a stable low-frequency squeezed vacuum field with periodically poled $KTiOPO_4$ at 1064 nm. Opt. Lett. **33**(2), 92 (2008)
38. Y. Takeno, M. Yukawa, H. Yonezawa, A. Furusawa, Observation of $-9$ dB quadrature squeezing with improvement of phase stability in homodyne measurement. Opt. Express **15**, 7 (2007)
39. H. Vahlbruch, M. Mehmet, S. Chelkowski, B. Hage, A. Franzen, N. Lastzka, S. Goßler, K. Danzmann, R. Schnabel, Observation of squeezed light with 10-dB quantum-noise reduction. Phys. Rev. Lett. **100**, 033602 (2008)
40. H. Yonezawa, K. Nagashima, A. Furusawa, Generation of squeezed light with a monolithic optical parametric oscillator: simultaneous achievement of phase matching and cavity resonance by temperature control. Opt. Exp. **18**(19), 20143–20150 (2010)
41. H. Vahlbruch, Squeezed light for gravitational wave astronomy. Ph.D. thesis, Albert Einstein Institute and Leibniz University of Hannover, Hannover, Germany, 2008
42. H. Vahlbruch, A. Khalaidovski, N. Lastzka, C. Gräf, K. Danzmann, R. Schnabel, The GEO600 squeezed light source. Class. Quantum Gravity **27**, 084027 (2010)
43. M. Mehmet, H. Vahlbruch, N. Lastzka, K. Danzmann, R. Schnabel, Observation of squeezed states with strong photon-number oscillations. Phys. Rev. A **81**, 013814 (2010)
44. T. Eberle, S. Steinlechner, J. Bauchrowitz, V. Händchen, H. Vahlbruch, M. Mehmet, H. Müller-Ebhardt, R. Schnabel, Quantum enhancement of the zero-area Sagnac interferometer topology for gravitational wave detection. Phys. Rev. Lett. **104**, 251102 (2010)
45. M. Mehmet, S. Ast, T. Eberle, S. Steinlechner, H. Vahlbruch, R. Schnabel, Squeezed light at 1550nm with a quantum noise reduction of 12.3 dB. Opt. Express **19**, 25 (2011)
46. A. Khalaidovski, H. Vahlbruch, N. Lastzka, C. Gräf, H. Lück, K. Danzmann, H. Grote, R. Schnabel, Status of the GEO600 squeezed-light laser (2011), arXiv:1112.0198v1
47. S.S.Y. Chua, M. Stefszky, C. Mow-Lowry, B. Buchler, S. Dwyer, D. Shaddock, P.K. Lam, D. McClelland, Backscatter tolerant squeezed light source for advanced gravitational-wave detectors. Opt. Lett. **36**(23), 4680 (2011)
48. M. Stefszky, C. Mow-Lowry, S. Chua, D. Shaddock, B. Buchler, H. Vahlbruch, A. Khalaidovski, R. Schnabel, P.K. Lam, D. McClelland, Balanced homodyne detection of optical quantum states at audio-band frequencies and below. Class. Quantum Gravity **29**, 145015 (2012)
49. S. Chua, K. McKenzie, B.C. Buchler, D.E. McClelland, Impact of non-stationary events on low frequency homodyne detection. J. Phys. Conf. Ser. **122**, 012023 (2008)
50. K. McKenzie, D.A. Shaddock, D.E. McClelland, B.C. Buchler, P.K. Lam, Experimental demonstration of a squeezing-enhanced power-recycled Michelson interferometer for gravitational wave detection. Phys. Rev. Lett. **88**(23), 231102 (2002)
51. H. Vahlbruch, S. Chelkowski, B. Hage, A. Franzen, K. Danzmann, R. Schnabel, Demonstration of a squeezed-light-enhanced power- and signal-recycled Michelson interferometer. Phys. Rev. Lett. **95**, 211102 (2005)
52. K. Goda, O. Miyakawa, E.E. Mikhailov, S. Saraf, R. Adhikari, K. McKenzie, R. Ward, S. Vass, A.J. Weinstein, N. Mavalvala, A quantum-enhanced prototype gravitational-wave detector. Nat. Phys. **4**, 472–476 (2008)

53. The LIGO Scientific Collaboration, A gravitational wave observatory operating beyond the quantum shot-noise limit. Nat. Phys. **7**, 962–965 (2011)
54. M. Pitkin, S. Reid, S. Rowan, J. Hough, Gravitational wave detection by interferometry (ground and space). Living Rev. Relativ. **14**(5) (2011)
55. D.F. Walls, G. Milburn, *Quantum Optics*, 2nd edn. (Springer, Berlin, 2008)
56. A. Buonanno, Y. Chen, Quantum noise in second generation, signal-recycled laser interferometric gravitational-wave detectors. Phys. Rev. D **64**, 042006 (2001)
57. A. Buonanno, Y. Chen, Improving the sensitivity to gravitational-wave sources by modifying the input-output optics of advanced interferometers. Phys. Rev. D **69**, 102004 (2004)

# Part I
# Background

# Chapter 2
# Gravitational Waves and the Quest for Their Direct Detection

**Abstract** This chapter introduces gravitational waves and reviews the optical-interferometric experiments for their direct detection. This is intended as an overview of the field—for further detail, a recommended source is Pitkin et al. (Living Rev Relativ 14(5):75 pp, 2011, [1]). Section 2.1 presents the basics of gravitational waves and potential sources of such waves in our Universe. Section 2.2 presents an introduction to current, second generation and future ground-based interferometric detectors. Section 2.3 provides an overview of noise sources currently affecting ground based measurements.

## 2.1 The Nature of Gravitational Waves

As presented in Einstein's theory of General Relativity [2], the gravitational force exists due to the curvature of space-time. This curvature implies that space-time is a flexible medium, and allows the existence of propagating waves of space-time known as *gravitational waves*. Gravitational waves are oscillations of space-time itself. This is intrinsically different to the more familiar electromagnetic waves, which propagate through space-time [3].

Some of the predicted astrophysical sources of gravitational waves include:

- **Inspiral/Coalescing Binary Systems** [4, 5]—Binary systems of compact objects, such as neutron stars and black holes, emit gravitational waves as they orbit. As angular momentum and energy are lost via gravitational waves, the orbital distance decays. Asymptoting to coalescence, the emitted gravitational waves increase in both strength and frequency, resulting in a chirp signal.
- **Spinning Massive Objects** [4]—Spinning massive objects will lose angular energy via emission of periodic gravitational waves, whose signal strength increases as the degree of axial-asymmetry increases. Possible candidates include non-symmetric pulsars and neutron stars.



S.S.Y. Chua, *Quantum Enhancement of a 4 km Laser Interferometer Gravitational-Wave Detector*, Springer Theses,
DOI 10.1007/978-3-319-17686-4_2

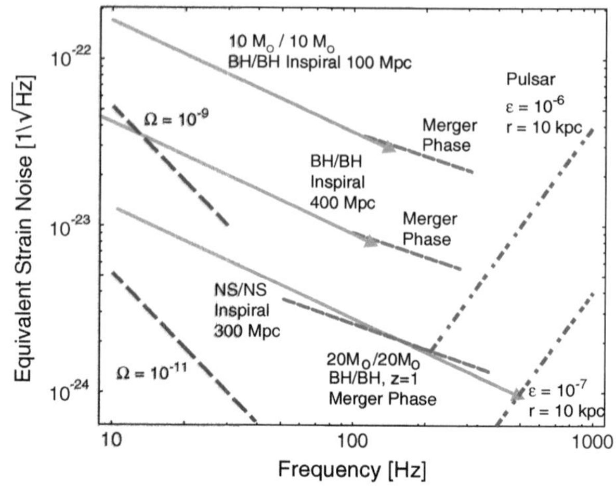


**Fig. 2.1** Predicted sources of gravitational waves—adapted from [4]. Symbol meanings: $M_O$ solar mass, $\varepsilon$ ellipticity, $\Omega$ stochastic background energy density

- **Stochastic Background** [4, 6]—A stochastic background of gravitational waves, similar to the cosmic microwave background, is expected to be a remnant from the early Universe. Predictions of signal strength for inspiral systems, spinning massive objects and Stochastic background are shown in Fig. 2.1.
- **Supernovae** [7, 8]—The non-symmetric dynamics of these stellar explosions are predicted to emit gravitational waves. Detecting these emissions will be a potential source of direct information about these events that are currently not well understood.

Propagating at the speed of light, gravitational waves interact very weakly with matter. This characteristic means that detecting gravitational waves may offer a relatively unhindered view of the Universe, a view that is inaccessible to electromagnetic radiation-based astronomy. For example, it allows the possibility of gaining insights from the formation of the Universe from $10^{-36}$ s after the Big Bang, or approximately 300,000 years earlier than electromagnetic-based astronomy is capable of [6], as well as dynamics of Supernovae core collapses [8] (above). Furthermore, gravitational waves are generated from these phenomena directly, thus carry direct information about the characteristics of these objects/events [9].

The lowest mode of gravitational-wave oscillation is the quadrapole, and can be best described by their effect on a region of space-time [10]. Figure 2.2 shows the effect of the two basic polarisations ($h_+$ and $h_\times$) of a passing gravitational wave on two rings of free-falling test particles. The rings are truncated and elongated in orthogonal directions.

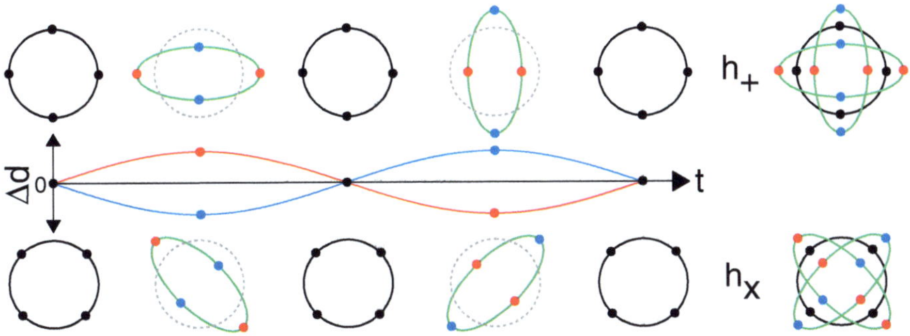


**Fig. 2.2** The effect of the two basic gravitational wave polarisations ($h_+$ and $h_\times$) passing into the page on free-falling test particles, where $\Delta$d represents the displacement of the particles from their neutral position over a full cycle. The strain is an exaggerated $h = 0.4$ to clearly demonstrate the effect

The strength of the gravitational wave is measured by the *strain*, $h$, or the fractional length change it induces, given by:

$$h = \frac{\delta L}{L} \tag{2.1}$$

where $\delta L$ is the change in length and $L$ is the original length that the change is measured over. The largest astrophysical sources/events are expected to have extraordinarily small strains, of order $h \approx 10^{-21}$ and lower [10].

Observations by Hulse and Taylor [11, 12] on a binary neutron star-pulsar system, PSR B1913+16, provided the indirect evidence of the existence of gravitational waves. They found that the changing orbital period of the binary system was consistent with the predicted loss of energy from system via the emission of gravitational waves [13].

## 2.2 Ground-Based Gravitational-Wave Interferometric Detectors

This section presents an overview of current and future ground-based interferometric gravitational-wave detector projects around the world. Other direct detection efforts and methods are being developed, such as space-based detectors [14, 15], resonant mass detectors [16–19], as well as pulsar timing arrays [20–24]. These efforts and methods are acknowledged but are not discussed.

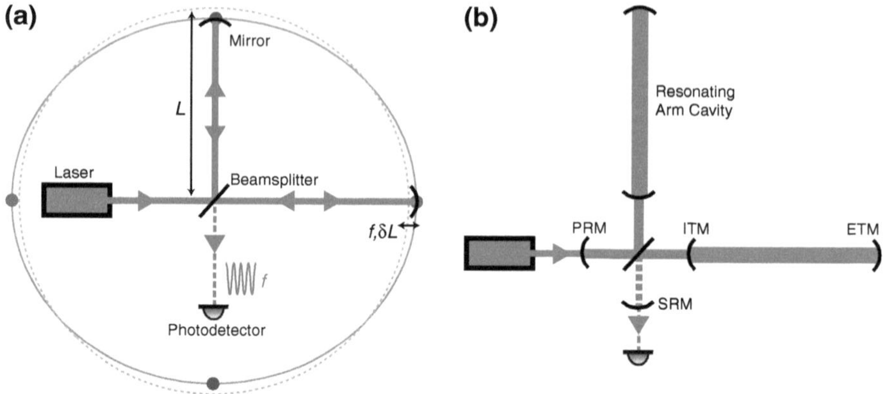


**Fig. 2.3** **a** A basic Michelson interferometer, showing the effect of a gravitational wave passing into the plane of the page. Details found in Sect. 2.2.1, **b** A Michelson interferometer with resonating arm cavities, power recycling and signal recycling mirrors. *PRM* Power Recycling Mirror, *ITM* Input Test Mass, *ETM* End Test Mass, *SRM* Signal Recycling Mirror, $f$ gravitational-wave frequency, $\delta L$ change in Michelson arm length

### *2.2.1 Interferometers for Direct Detection*

Ground-based interferometric gravitational-wave detectors are based on a Michelson Interferometer topology. For the remainder of this thesis, references to "gravitational-wave detectors" will specifically refer to such devices. A basic Michelson Interferometer is shown in Fig. 2.3a. Continuous-wave light from a laser is split into two beams with a beamsplitter. The two beams then travel in the orthogonal *arms* of the Michelson (length $L$). Both beams are reflected back towards and interfered on the beamsplitter via the *test mass* mirrors, resulting in an interference signal that encodes information about the relative optical path difference between the two arms, detectable by the photodetector. Any change in the relative optical path difference ($\delta L$) of the two arms will change the interference signal at the beamsplitter. A gravitational wave of a particular Fourier frequency $f$ reveals itself in an interferometer as a modulation signal on the light at the same frequency.

Gravitational-wave detectors use additional techniques to further increase the sensitivity. These include resonating optical cavities for the Michelson arms, optical recycling mirrors, as shown in Fig. 2.3b. The first generation gravitational-wave detectors included LIGO [25], Virgo [26], GEO600 [27] and TAMA300 [28] interferometers. All first generation detectors broadly reached their design sensitivities by 2006.

### 2.2.2 The "Generation 1.5" Detectors

The "Generation 1.5" interferometers formed intermediate upgrade stages for LIGO and Virgo instruments, working towards their respective second generation devices.

- **Enhanced LIGO** [31]—The Enhanced Laser Interferometer Gravitational-wave Observatory (also known as eLIGO) were two Michelson interferometers, operated by the US-led LIGO Scientific Collaboration. Situated at two sites in the United States (Hanford, Washington State and Livingston, Louisiana), they are purposely separated by approximately 3000 km to allow for triangulation of gravitational-wave sources in the sky. Each site had an interferometer with 4 km arm cavities and power recycling.
- **Virgo+** [32]—The Virgo+ Interferometer was also a power-recycled Michelson interferometer, but with 3 km resonating arm cavities. Situated near Pisa in Italy and operated by the French-Italian Virgo Collaboration, Virgo+ had smaller peak-strain sensitivity due to shorter arm lengths, but deployed more advanced seismic isolation systems resulting in better sensitivity at lower detection frequencies.

A joint scientific measurement run between Enhanced LIGO and Virgo+ [33] was concluded in October 2010. The start of the second-generation upgrade/installation programs for Enhanced LIGO and Virgo+ then immediately followed. The estimated development timelines from Generation 1.5 to second generation detectors are shown in Fig. 2.5. The work presented in the later chapters of this thesis were undertaken on the Enhanced LIGO interferometer at Hanford, thus greater detail about the Enhanced LIGO interferometer will be covered in Chap. 8.

### 2.2.3 Second Generation Detectors

The locations of the gravitational-wave detectors comprising the second generation network described below, are shown in Fig. 2.4a.

- **Advanced LIGO** [34]—The Advanced LIGO interferometer network will consist of three 4 km interferometers, one interferometer located in each of the two USA sites and the third interferometer located in India (LIGO India with the IndIGO consortium [35]). These interferometers will have arm cavities and use *dual recycling*, that is both power recycling and signal recycling. With improved seismic isolation, test mass mirrors, mirror coatings and mirror suspension technologies, and an increase in operating laser power, the target sensitivity improvement is an order of 10 in magnitude over (initial) LIGO. This translates to an increased observational reach of 1000 in volume space compared to LIGO, pictorially shown in Fig. 2.4b.
- **Advanced Virgo** [36]—The Advanced Virgo interferometer will be a 3 km dual-recycled interferometer with arm cavities, with increased operating laser power and improved mirrors and mirror coatings to augment their advanced seismic isolation

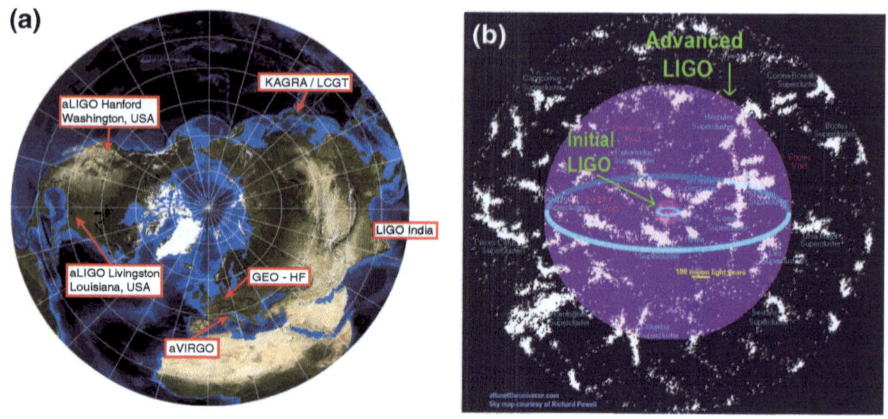


**Fig. 2.4** **a** Second-generation gravitational-wave interferometric detector sites around the world; **b** Observational reach of Advanced LIGO, one thousand times larger volume space compared to Initial LIGO [29]. Sky map courtesy of Richard Powell [30]

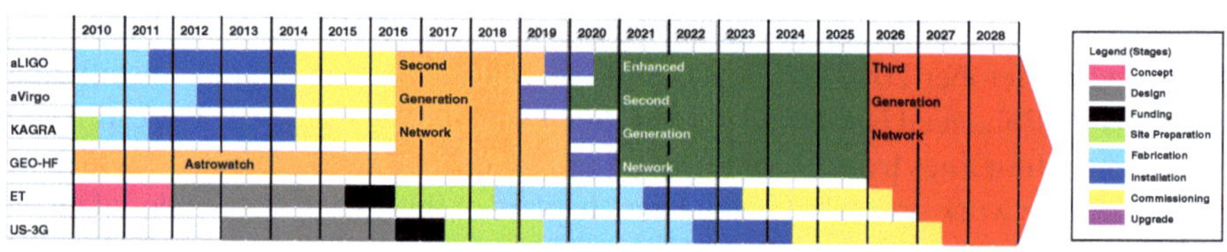


**Fig. 2.5** Development timelines for ground-based gravitational-wave detectors—adapted from [40]

system. The target sensitivity improvement is also an order of 10 in magnitude relative to first generation Virgo.

- **KAGRA (LCGT)** [37]—Located underground at Kamioka, Gifu Prefecture, Japan, construction of KAGRA Large-scale Cryogenic Gravitational-wave Telescope (LCGT) began in January 2012. KAGRA will be a 3 km power-recycled interferometer with arm cavities, and will be the first underground kilometre-scale interferometer and the first to employ cryogenics as part of its baseline technology.
- **GEO-HF/GEO600** [38]—Operated by the British-German GEO collaboration, GEO-HF is the second generation upgrade of GEO600 [27] interferometer, whose main upgrade took place during the S6/VSR2-3 Science Run. The sole interferometer in operation during the second generation detector upgrade/installation phase (labelled 'Astrowatch' in Fig. 2.5), it is in dual-recycled configuration with 600 m folded arms and no arm cavities. It is the first interferometer to employ squeezed states [39] as part of its baseline operation.

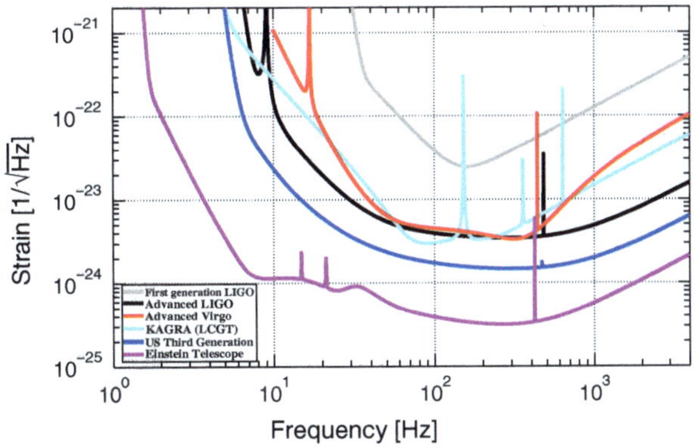


**Fig. 2.6** Design strain-sensitivities for selected second generation and future laser-interferometric gravitational wave detectors. The achieved design sensitivity for first generation LIGO [25], is also plotted

### 2.2.4 Third Generation Detectors

- **Einstein Telescope** [41, 42]—Design studies for a third generation European detector called the Einstein Telescope have been undertaken. This interferometer looks to combine dual-recycled Michelsons with arm cavities, squeezed states and cryogenics, along with being housed within a deep-underground facility.
- **Third Generation LIGO** [43]—The early stages of planning/discussion for the third generation LIGO instruments started in early 2012. Various designs with different technologies and optical topologies are being studied, including different materials for test masses and different cooling/cryogenic strategies. However, one common technology included in all third generation LIGO designs being discussed is the use of squeezed states.

### 2.2.5 Strain Sensitivities and Development Timelines

For the second generation and third generation detectors, their estimated development timelines are shown in Fig. 2.5, and their respective design/predicted sensitivities have been graphically presented in Fig. 2.6.

## 2.3 Noise Sources Affecting Ground-Based Interferometers

For the second generation interferometers, the target strain sensitivity is of order $h = 10^{-23}/\sqrt{\text{Hz}}$, as shown in Fig. 2.6. This presents a great challenge within the audio frequency detection band of interest (10 Hz–10 kHz). A multitude of noise sources can limit the strain sensitivity, some of which are introduced below. Their contributions to the Advanced LIGO sensitivity, as an example, are shown in Fig. 2.7.

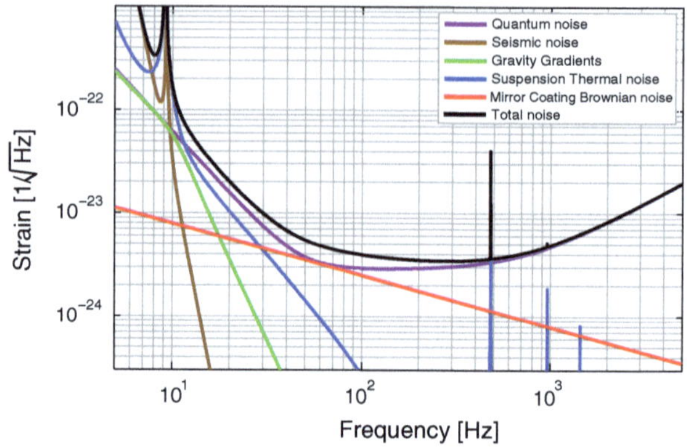


**Fig. 2.7** The predicted noise budget of Advanced LIGO, showing the contributions of the various limiting noise sources. Plot made using GWINC v3 [44]

### *2.3.1 Quantum Noise*

Quantum noise in interferometric detectors arise due to the quantum mechanical fluctuations of the electromagnetic field used to sense the displacement of the test masses. This noise source manifests through both measurement photon arrival time at the photodetector and photon momentum transfer onto the test masses, and is related to the operating optical laser power. Quantum noise is expected to be the main sensitivity-limiting noise source in second generation gravitational-wave detectors across the audio frequency detection band.

The primary focus of the work in this thesis is the generation and application of squeezed states to surpass quantum noise to enhance interferometric detector measurements. Quantum noise will be in detail in subsequent chapters.

### *2.3.2 Thermal Noise*

The term 'thermal noise' covers the displacement noise of the test masses that arise from the thermal fluctuations of the atoms that make up the components. This includes the thermal noise from mirror coatings [45], surface vibration modes of the test mass mirrors [46], internal modes of the test mass mirrors [47], thermal noise from the suspension wires, and thermo-refractive noise [48] through input-test-mass transmission.

The current approach for mitigating thermal noise is to use low mechanical loss materials for test masses, optics and suspensions [10]. However, the choice of such materials is limited, complicated by key requirements such as their light absorption properties. Other approaches being considered include cryogenic test masses [49, 50] and exotic test mass surface structures that reduce the need for mirror coatings [51].

### 2.3.3 Seismic Noise

To reach the required sensitivity, the test masses must be isolated from seismic ground motion. The displacement spectral density of ground motion at the LIGO sites are of the order $10^{-8}\text{m}/\sqrt{\text{Hz}}$ at 1 Hz, decreasing gradually with increasing Fourier frequency [52]. A combination of active and passive isolation is typically deployed, however, it remains a limiting floor at lower frequencies of the audio detection band. The underground location of the KAGRA detector was chosen to minimise the impact of seismic noise.

### 2.3.4 Gravity Gradient Noise

Fluctuations in the gravitational field of the local environment cause uncorrelated displacement noise on to the test masses. This noise, called gravity gradient noise, is caused by moving mass bodies (e.g. trains, cars, people), and changes in environmental-media densities (e.g. atmospheric pressure, subterranean sediment settlement) [53, 54]. The Einstein Telescope looks toward deep-underground facilities to mitigate the effect of this noise source. This noise source represents the fundamental low frequency (to a few Hz) limit of ground-based interferometers.

### 2.3.5 Other Noise Sources

Other noise sources affecting the sensitivity of gravitational-wave detectors include feedback control noise, electronic component noise, photothermal noise [55], residual gas [10, 56] and stray light [57–59]. These noise sources are not expected to limit detection-sensitivity.

## 2.4 Summary

An overview of gravitational waves and the quest for their direct detection has been presented. The nature of gravitational waves, and astrophysical sources predicted to emit gravitational waves were reviewed. Generation 1.5 and second generation ground-based interferometric detectors were then introduced. The sensitivity-limiting noise sources affecting gravitational-wave detector measurement were covered. Quantum noise of the electromagnetic field broadly limits the detection sensitivity across the audio gravitational-wave detection band.

## References


1. M. Pitkin, S. Reid, S. Rowan, J. Hough, Gravitational wave detection by interferometry (ground and space). Living Rev. Relativ. **14**(5), 75 (2011)
2. A. Einstein, The foundation of the general theory of relativity. Ann. Phys. Lpz **49**, 769 (1916)
3. K.S. Thorne, *Gravitational Waves* (1995), arXiv:gr-qc/9506086v1
4. C. Cutler, K.S. Thorne, *An Overview of Gravitational-Waves Sources* (2002), arXiv:gr-qc/0204090v1
5. The LIGO Scientific Collaboration and the Virgo Collaboration, J. Abadie et al., Predictions for the rates of compact binary coalescences observable by ground-based gravitational-wave detectors. Class. Quantum Gravity **27**, 173001 (2010)
6. M. Maggiore, Gravitational wave experiments and early universe cosmology. Phys. Rep. **331**, 283 (2000)
7. C.D. Ott, Probing the core-collapse supernova mechanism with gravitational waves. Class. Quantum Gravity **26**, 204015 (2009)
8. J. Logue, C.D. Ott, I.S. Heng, P. Kalmus, J.H.C. Scargill, *Inferring Core-Collapse Supernova Physics with Gravitational Waves* (2012), arXiv:1202.3256v2
9. E.E. Flanagan, S.A. Hughes, Measuring gravitational waves from binary black hole coalescences. I. Signal to noise for inspiral, merger, and ringdown. Phys. Rev. D **57**, 4535 (1998)
10. P.R. Saulson, *Fundamentals of Interferometric Gravitational Wave Detectors* (World Scientific, Singapore, 1994)
11. R.A. Hulse, J.H. Taylor, Discovery of a pulsar in a binary system. Astrophys. J. (Lett.) **195**, L51 (1975)
12. J.H. Taylor, J.M. Weisberg, Further experimental tests of relativistic gravity using the binary pulsar PSR 1913+16. Astrophys. J. **345**, 434 (1989)
13. J.H. Taylor, J.M. Weisberg, A new test of general relativity: gravitational radiation and the binary pulsar PSR 1913+16. Astrophys. J. **253**, 908 (1982)
14. P. Amaro-Seoane, S. Aoudia, S. Babak, P. Binétruy, E. Berti, A. Bohé, C. Caprini, M. Colpi, N.J. Cornish, K. Danzmann, J.-F. Dufaux, J. Gair, O. Jennrich, P. Jetzer, A. Klein, R.N. Lang, A. Lobo, T. Littenberg, S.T. McWilliams, G. Nelemans, A. Petiteau, E.K. Porter, B.F. Schutz, A. Sesana, R. Stebbins, T. Sumner, M. Vallisneri, S. Vitale, M. Volonteri, H. Ward, *Doing Science with eLISA: Astrophysics and Cosmology in the Millihertz Regime* (2012), arXiv:1201.3621v1
15. S. Kawamura, M. Ando, N. Seto, S. Sato, T. Nakamura, K. Tsubono, N. Kanda, T. Tanaka, J. Yokoyama, I. Funaki, K. Numata, K. Ioka, T. Takashima, K. Agatsuma, T. Akutsu, K. Aoyanagi, K. Arai, A. Araya, H. Asada, Y. Aso, D. Chen, T. Chiba, T. Ebisuzaki, Y. Ejiri, M. Enoki, Y. Eriguchi, M.-K. Fujimoto, R. Fujita, M. Fukushima, T. Futamase, T. Harada, T. Hashimoto, K. Hayama, W. Hikida, Y. Himemoto, H. Hirabayashi, T. Hiramatsu, F.-L. Hong, H. Horisawa, M. Hosokawa, K. Ichiki, T. Ikegami, K.T. Inoue, K. Ishidoshiro, H. Ishihara, T. Ishikawa, H. Ishizaki, H. Ito, Y. Itoh, K. Izumi, I. Kawano, N. Kawashima, F. Kawazoe, N. Kishimoto, K. Kiuchi, S. Kobayashi, K. Kohri, H. Koizumi, Y. Kojima, K. Kokeyama, W. Kokuyama, K. Kotake, Y. Kozai, H. Kunimori, H. Kuninaka, K. Kuroda, S. Kuroyanagi, K. Maeda, H. Matsuhara, N. Matsumoto, Y. Michimura, O. Miyakawa, U. Miyamoto, S. Miyoki, M.Y. Morimoto, T. Morisawa, S. Moriwaki, S. Mukohyama, M. Musha, S. Nagano, I. Naito, K. Nakamura, H. Nakano, K. Nakao, S. Nakasuka, Y. Nakayama, K. Nakazawa, E. Nishida, K. Nishiyama, A. Nishizawa, Y. Niwa, T. Noumi, Y. Obuchi, M. Ohashi, N. Ohishi, M. Ohkawa, K. Okada, N. Okada, K. Oohara, N. Sago, M. Saijo, R. Saito, M. Sakagami, S. Sakai, S. Sakata, M. Sasaki, T. Sato, M. Shibata, H. Shinkai, A. Shoda, K. Somiya, H. Sotani, N. Sugiyama, Y. Suwa, R. Suzuki, H. Tagoshi, F. Takahashi, K. Takahashi, K. Takahashi, R. Takahashi, R. Takahashi, T. Takahashi, H. Takahashi, T. Akiteru, T. Takano, N. Tanaka, K. Taniguchi, A. Taruya, H. Tashiro, Y. Torii, M. Toyoshima, S. Tsujikawa, Y. Tsunesada, A. Ueda, K. Ueda, M. Utashima, Y. Wakabayashi, K. Yagi, H. Yamakawa, K. Yamamoto, T. Yamazaki, C.-M. Yoo, S. Yoshida, T. Yoshino, K.-X. Sun, The Japanese space gravitational wave antenna: DECIGO. Class. Quantum Gravity **28**(9), 094011 (2011)

16. L. Gottardi, A. de Waard, O. Usenko, G. Frossati, Sensitivity of the spherical gravitational wave detector MiniGRAIL operating at 5 K. Phys. Rev. D **76**, 102005 (2007)
17. P. Astone, R. Ballantini, D. Babusci, M. Bassan, P. Bonifazi, G. Cavallari, A. Chincarini, E. Coccia, S. D'Antonio, M. Di Paolo Emilio, V. Fafone, S. Foffa, G. Gemme, G. Giordano, M. Maggiore, A. Marini, Y. Minenkov, I. Modena, G. Modestino, A. Moleti, G.V. Pallottino, R. Parodi, G. Pizzella, EXPLORER and NAUTILUS gravitational wave detectors: a status report. Class. Quantum Gravity **25**, 114048 (2008)
18. A. Vinante, (for the AURIGA Collaboration), Present performance and future upgrades of the AURIGA capacitive readout. Class. Quantum Gravity **23**, S103–S110 (2006)
19. O.D. Aguiar, L.A. Andrade, J.J. Barroso, P.J. Castro, C.A. Costa, S.T. de Souza, A. de Waard, A.C. Fauth, C. Frajuca, G. Frossati, S.R. Furtado, X. Gratens, T.M.A. Maffei, N.S. Magalhaes, R. M. Marinho Jr., N.F. Oliveira Jr., G.L. Pimentel, M.A. Remy, M.E. Tobar, E. Abdalla, M.E.S. Alves, D.F.A. Bessada, F.S. Bortoli, C.S.S. Brandao, K.M.F. Costa, H.A.B. de Araujo, J.C.N. de Araujo, E.M. de Gouveia Dal Pino, W. de Paula, E.C. de Rey Neto, E.F.D. Evangelista, C.H. Lenzi, G.F. Marranghello, O.D. Miranda, S.R. Oliveira, R. Opher, E.S. Pereira, C. Stellati, J. Weber, The Schenberg spherical gravitational wave detector: the first commissioning runs. Class. Quantum Gravity **25**, 114042 (2008)
20. G. Hobbs, A. Archibald, Z. Arzoumanian, D. Backer, M. Bailes, N.D.R. Bhat, M. Burgay, S. Burke-Spolaor, D. Champion, I. Cognard, W. Coles, J. Cordes, P. Demorest, G. Desvignes, R.D. Ferdman, L. Finn, P. Freire, M. Gonzalez, J. Hessels, A. Hotan, G. Janssen, F. Jenet, A. Jessner, C. Jordan, V. Kaspi, M. Kramer, V. Kondratiev, J. Lazio, K. Lazaridis, K.J. Lee, Y. Levin, A. Lommen, D. Lorimer, R. Lynch, A. Lyne, R. Manchester, M. McLaughlin, D. Nice, S. Oslowski, M. Pilia, A. Possenti, M. Purver, S. Ransom, J. Reynolds, S. Sanidas, J. Sarkissian, A. Sesana, R. Shannon, X. Siemens, I. Stairs, B. Stappers, D. Stinebring, G. Theureau, R. van Haasteren, W. van Straten, J.P.W. Verbiest, D.R.B. Yardley, X.P. You, The international pulsar timing array project: using pulsars as a gravitational wave detector. Class. Quantum Gravity **27**, 084013 (2010)
21. G.B. Hobbs, M. Bailes, N.D.R. Bhat, S. Burke-Spolaor, D.J. Champion, W. Coles, A. Hotan, F. Jenet, L. Kedziora-Chudczer, J. Khoo, K.J. Lee, A. Lommen, R.N. Manchester, J. Reynolds, J. Sarkissian, W. van Straten, S. To, J.P.W. Verbiest, D. Yardley, X.P. You, Gravitational-wave detection using pulsars: status of the Parkes pulsar timing array project. Publ. Astron. Soc. Aust. **26**, 103 (2009)
22. G.H. Janssen, B.W. Stappers, M. Kramer, M. Purver, A. Jessner, I. Cognard, European pulsar timing array. AIP Conf. Proc. **983**, 633–635 (2008)
23. F. Jenet, L.S. Finn, J. Lazio, A. Lommen, M. McLaughlin, I. Stairs, D. Stinebring, J. Verbiest. *The North American Nanohertz Observatory for Gravitational Waves* (2009), arXiv:0909.1058
24. P.E. Dewdney, P.J. Hall, R.T. Schilizzi, T.J.L.W. Lazio, The square kilometre array. Proc. IEEE **97**(8), 1482–1496 (2009)
25. The LIGO Scientific Collaboration (B.P. Abbott et al.), LIGO: the Laser Interferometer Gravitational-Wave Observatory. Rep. Prog. Phys. **72**, 076901 (2009)
26. T. Accadia, F. Acernese, M. Alshourbagy, P. Amico, F. Antonucci, S. Aoudia, N. Arnaud, C. Arnault, K.G. Arun, P. Astone, S. Avino, D. Babusci, G. Ballardin, F. Barone, G. Barrand, L. Barsotti, M. Barsuglia, A. Basti, T.S. Bauer, F. Beauville, M. Bebronne, M. Bejger, M.G. Beker, F. Bellachia, A. Belletoile, J.L. Beney, M. Bernardini, S. Bigotta, R. Bilhaut, S. Birindelli, M. Bitossi, M.A. Bizouard, M. Blom, C. Boccara, D. Boget, F. Bondu, L. Bonelli, R. Bonnand, V. Boschi, L. Bosi, T. Bouedo, B. Bouhou, A. Bozzi, L. Bracci, S. Braccini, C. Bradaschia, M. Branchesi, T. Briant, A. Brillet, V. Brisson, L. Brocco, T. Bulik, H.J. Bulten, D. Buskulic, C. Buy, G. Cagnoli, G. Calamai, E. Calloni, E. Campagna, B. Canuel, F. Carbognani, L. Carbone, F. Cavalier, R. Cavalieri, R. Cecchi, G. Cella, E. Cesarini, E. Chassande-Mottin, S. Chatterji, R. Chiche, A. Chincarini, A. Chiummo, N. Christensen, A.C. Clapson, F. Cleva, E. Coccia, P.-F. Cohadon, C.N. Colacino, J. Colas, A. Colla, M. Colombini, G. Conforto, A. Corsi, S. Cortese, F. Cottone, J.-P. Coulon, E. Cuoco, S. D'Antonio, G. Daguin, A. Dari, V. Dattilo, P.Y. David, M. Davier, R. Day, G. Debreczeni, G. De Carolis, M. Dehamme, R. Del Fabbro, W. Del Pozzo, M. del Prete, L. Derome, R. De Rosa, R. DeSalvo, M. Dialinas, L. Di Fiore, A. Di Lieto, M.

Di Paolo Emilio, A. Di Virgilio, A. Dietz, M. Doets, P. Dominici, A. Dominjon, M. Drago, C. Drezen, B. Dujardin, B. Dulach, C. Eder, A. Eleuteri, D. Enard, M. Evans, L. Fabbroni, V. Fafone, H. Fang, I. Ferrante, F. Fidecaro, I. Fiori, R. Flaminio, D. Forest, L.A. Forte, J.-D. Fournier, L. Fournier, J. Franc, O. Francois, S. Frasca, F. Frasconi, A. Freise, A. Gaddi, M. Galimberti, L. Gammaitoni, P. Ganau, C. Garnier, F. Garufi, M.E. Gáspár, G. Gemme, E. Genin, A. Gennai, G. Gennaro, L. Giacobone, A. Giazotto, G. Giordano, L. Giordano, C. Girard, R. Gouaty, A. Grado, M. Granata, V. Granata, X. Grave, C. Greverie, H. Groenstege, G.M. Guidi, S. Hamdani, J.-F. Hayau, S. Hebri, A. Heidmann, H. Heitmann, P. Hello, G. Hemming, E. Hennes, R. Hermel, P. Heusse, L. Holloway, D. Huet, M. Iannarelli, P. Jaranowski, D. Jehanno, L. Journet, S. Karkar, T. Ketel, H. Voet, J. Kovalik, I. Kowalska, S. Kreckelbergh, A. Krolak, J.C. Lacotte, B. Lagrange, P. La Penna, M. Laval, J.C. Le Marec, N. Leroy, N. Letendre, T.G.F. Li, B. Lieunard, N. Liguori, O. Lodygensky, B. Lopez, M. Lorenzini, V. Loriette, G. Losurdo, M. Loupias, J.M. Mackowski, T. Maiani, E. Majorana, C. Magazzù, I. Maksimovic, V. Malvezzi, N. Man, S. Mancini, B. Mansoux., M. Mantovani, F. Marchesoni, F. Marion, P. Marin, J. Marque, F. Martelli, A. Masserot, L. Massonnet, G. Matone, L. Matone, M. Mazzoni, F. Menzinger, C. Michel, L. Milano, Y. Minenkov, S. Mitra, M. Mohan, J.-L. Montorio, R. Morand, F. Moreau, J. Moreau, N. Morgado, A. Morgia, S. Mosca, V. Moscatelli, B. Mours, P. Mugnier, F.-A. Mul, L. Naticchioni, I. Neri, F. Nocera, E. Pacaud, G. Pagliaroli, A. Pai, L. Palladino, C. Palomba, F. Paoletti, R. Paoletti, A. Paoli, S. Pardi, G. Parguez, M. Parisi, A. Pasqualetti, R. Passaquieti, D. Passuello, M. Perciballi, B. Perniola, G. Persichetti, S. Petit, M. Pichot, F. Piergiovanni, M. Pietka, R. Pignard, L. Pinard, R. Poggiani, P. Popolizio, T. Pradier, M. Prato, G. A. Prodi, M. Punturo, P. Puppo, K. Qipiani, O. Rabaste, D.S. Rabeling, I. Rácz, F. Raffaelli, P. Rapagnani, S. Rapisarda, V. Re, A. Reboux, T. Regimbau, V. Reita, A. Remilleux, F. Ricci, I. Ricciardi, F. Richard, M. Ripepe, F. Robinet, A. Rocchi, L. Rolland, R. Romano, D. Rosinska, P. Roudier, P. Ruggi, G. Russo, L. Salconi, V. Sannibale, B. Sassolas, D. Sentenac, S. Solimeno, R. Sottile, L. Sperandio, R. Stanga, R. Sturani, B. Swinkels, M. Tacca, R. Taddei, L. Taffarello, M. Tarallo, S. Tissot, A. Toncelli, M. Tonelli, O. Torre, E. Tournefier, F. Travasso, C. Tremola, E. Turri, G. Vajente, J.F.J. van den Brand, C. Van Den Broeck, S. van der Putten, M. Vasuth, M. Vavoulidis, G. Vedovato, D. Verkindt, F. Vetrano, O. Véziant, A. Vicerê, J.-Y. Vinet, S. Vilalte, S. Vitale, H. Vocca, R.L. Ward, M. Was, K. Yamamoto, M. Yvert, J.-P. Zendri, Z. Zhang, Virgo: a laser interferometer to detect gravitational waves. J. Instrum. **7**(03), P03012 (2012)

27. H. Grote (for the LIGO Scientific Collaboration), The GEO 600 status. Class. Quantum Gravity **27**, 084003 (2010)
28. M. Ando, K. Arai, R. Takahashi, G. Heinzel, S. Kawamura, D. Tatsumi, N. Kanda, H. Tagoshi, A. Araya, H. Asada, Y. Aso, M.A. Barton., M.-K. Fujimoto, M. Fukushima, T. Futamase, K. Hayama, G. Horikoshi, H. Ishizuka, N. Kamikubota, K. Kawabe, N. Kawashima, Y. Kobayashi, Y. Kojima, K. Kondo, Y. Kozai, K. Kuroda, N. Matsuda, N. Mio, K. Miura, O. Miyakawa, S.M. Miyama, S. Miyoki, S. Moriwaki, M. Musha, S. Nagano, K. Nakagawa, T. Nakamura, K. Nakao, K. Numata, Y. Ogawa, M. Ohashi, N. Ohishi, S. Okutomi, K. Oohara, S. Otsuka, Y. Saito, M. Sasaki, S. Sato, A. Sekiya, M. Shibata, K. Somiya, T. Suzuki, A. Takamori, T. Tanaka, S. Taniguchi, S. Telada, K. Tochikubo, T. Tomaru, K. Tsubono, N. Tsuda, T. Uchiyama, A. Ueda, K. Ueda, K. Waseda, Y. Watanabe, H. Yakura, K. Yamamoto, T. Yamazaki, (TAMA Collaboration). Stable operation of a 300-m laser interferometer with sufficient sensitivity to detect gravitational-wave events within our galaxy. Phys. Rev. Lett. **86**(18), 3950–3954 (2001)
29. Advanced LIGO Website—Summary, https://www.advancedligo.mit.edu/summary.html
30. Atlas of the universe, http://www.atlasoftheuniverse.com
31. R. Adhikari, P. Fritschel, S. Waldman, Enhanced LIGO—Technical Note LIGO-T060156-01-I (2006), http://www.ligo.caltech.edu/docs/T/T060156-01.pdf
32. T. Accadia, B.L. Swinkels (for the Virgo Collaboration), Commissioning status of the Virgo interferometer. Class. Quantum Gravity **27**, 084002 (2010)
33. The LIGO Scientific Collaboration and The Virgo Collaboration (J. Abadie et al.), *Sensitivity Achieved by the LIGO and Virgo Gravitational Wave Detectors during LIGO's Sixth and Virgo's Second and Third Science Runs* (2012), arXiv:1203.2674v2

34. G.M. Harry, (for the LIGO Scientific Collaboration), Advanced LIGO: the next generation of gravitational wave detectors. Class. Quantum Gravity **27**, 084006 (2010)
35. Indian Initiative in Gravitational-wave Observations—IndIGO, http://www.gw-indigo.org/tiki-index.php?page=Welcome
36. The Virgo Collaboration, Advanced Virgo Baseline Design—VIR-027A-09 (2009), https://tds.ego-gw.it/itf/tds/file.php?callFile=VIR-0027A-09.pdf
37. K. Kuroda (for the LCGT Collaboration), Status of LCGT. Class. Quantum Gravity **27**, 084004 (2010)
38. B. Willke, P. Ajith, B. Allen, P. Aufmuth, C. Aulbert, S. Babak, R. Balasubramanian, B.W. Barr, S. Berukoff, A. Bunkowski, G. Cagnoli, C.A. Cantley, M.M. Casey, S. Chelkowski, Y. Chen, D. Churches, T. Cokelaer, C.N. Colacino, D.R.M. Crooks, C. Cutler, K. Danzmann, R.J. Dupuis, E. Elliffe, C. Fallnich, A. Franzen, A. Freise, I. Gholami, S. Goßler, A. Grant, H. Grote, S. Grunewald, J. Harms, B. Hage, G. Heinzel, I.S. Heng, A. Hepstonstall, M. Heurs, M. Hewitson, S. Hild, J. Hough, Y. Itoh, G. Jones, R. Jones, S.H. Huttner, K. Kötter, B. Krishnan, P. Kwee, H. Lück, M. Luna, B. Machenschalk, M. Malec, R.A. Mercer, T. Meier, C. Messenger, S. Mohanty, K. Mossavi, S. Mukherjee, P. Murray, G.P. Newton, M.A. Papa, M. Perreur-Lloyd, M. Pitkin, M.V. Plissi, R. Prix, V. Quetschke, V. Re, T. Regimbau, H. Rehbein, S. Reid, L. Ribichini, D.I. Robertson, N.A. Robertson, C. Robinson, J.D. Romano, S. Rowan, A. Rüdiger, B.S. Sathyaprakash, R. Schilling, R. Schnabel, B.F. Schutz, F. Seifert, A.M. Sintes, J.R. Smith, P.H. Sneddon, K.A. Strain, I. Taylor, R. Taylor, A. Thüring, C. Ungarelli, H. Vahlbruch, A. Vecchio, J. Veitch, H. Ward, U. Weiland, H. Welling, L. Wen, P. Williams, W. Winkler, G. Woan, R. Zhu, The GEO-HF project. Class. Quantum Gravity **23**, S207 (2006)
39. H. Vahlbruch, A. Khalaidovski, N. Lastzka, C. Gräf, K. Danzmann, R. Schnabel, The GEO600 squeezed light source. Class. Quantum Gravity **27**, 084027 (2010)
40. Gravitational Wave International Committee (GWIC), GWIC Roadmap—The Future of Gravitational Wave Astronomy, A Global Plan, https://gwic.ligo.org/roadmap/
41. S. Hild, M. Abernathy, F. Acernese, P. Amaro-Seoane, N. Andersson, K. Arun, F. Barone, B. Barr, M. Barsuglia, M. Beker, N. Beveridge, S. Birindelli, S. Bose, L. Bosi, S. Braccini, C. Bradaschia, T. Bulik, E. Calloni, G. Cella, E. Chassande, Mottin, S. Chelkowski, A. Chincarini, J. Clark, E. Coccia, C. Colacino, J. Colas, A. Cumming, L. Cunningham, E. Cuoco, S. Danilishin, K. Danzmann, R. De Salvo, T. Dent, R. De Rosa, L. Di Fiore, A. Di Virgilio, M. Doets, V. Fafone, P. Falferi, R. Flaminio, J. Franc, F. Frasconi, A. Freise, D. Friedrich, P. Fulda, J. Gair, G. Gemme, E. Genin, A. Gennai, A. Giazotto, K. Glampedakis, C. Gräf, M. Granata, H. Grote, G. Guidi, A. Gurkovsky, G. Hammond, M. Hannam, J. Harms, D. Heinert, M. Hendry, I. Heng, E. Hennes, J. Hough, S. Husa, S. Huttner, G. Jones, F. Khalili, K. Kokeyama, K. Kokkotas, B. Krishnan, T.G.F. Li, M. Lorenzini, H. Lück, E. Majorana, I. Mandel, V. Mandic, M. Mantovani, I. Martin, C. Michel, Y. Minenkov, N. Morgado, S. Mosca, B. Mours, H. Müller-Ebhardt, P. Murray, R. Nawrodt, J. Nelson, R. O'shaughnessy, C.D. Ott, C. Palomba, A. Paoli, G. Parguez, A. Pasqualetti, R. Passaquieti, D. Passuello, L. Pinard, W. Plastino, R. Poggiani, P. Popolizio, M. Prato, M. Punturo, P. Puppo, D. Rabeling, P. Rapagnani, J. Read, T. Regimbau, H. Rehbein, S. Reid, F. Ricci, F. Richard, A. Rocchi, S. Rowan, A. Rüdiger, L. Santamaría, B. Sassolas, B. Sathyaprakash, R. Schnabel, C. Schwarz, P. Seidel, A. Sintes, K. Somiya, F. Speirits, K. Strain, S. Strigin, P. Sutton, S. Tarabrin, A Thüring, J. van den Brand, M. van Veggel, C. van den Broeck, A. Vecchio, J. Veitch, F. Vetrano, A. Vicere, S. Vyatchanin, B. Willke, G. Woan, K. Yamamoto, Sensitivity studies for third-generation gravitational wave observatories. Class. Quantum Gravity **28**, 094013 (2011)
42. S. Hild, S. Chelkowski, A. Freise, J. Franc, N. Morgado, R. Flaminio, R. DeSalvo, A xylophone configuration for a third-generation gravitational wave detector. Class. Quantum Gravity **27**, 015003 (2010)
43. R. Adhikari, K. Arai, S. Ballmer, E. Gustafson, S. Hild, Report of the 3rd generation LIGO detector strawman workshop (2012). LIGO-T1200031-v3
44. Gravitational Wave Interferometer Noise Calculator (GWINC) v3, https://awiki.ligo-wa.caltech.edu/aLIGO/GWINC

45. G.M. Harry, A.M. Gretarsson, P.R. Saulson, S.E. Kittelberger, S.D. Penn, W.J. Startin, S. Rowan, M.M. Fejer, D.R.M. Crooks, G. Cagnoli, J. Hough, N. Nakagawa, Thermal noise in interferometric gravitational wave detectors due to dielectric optical coatings. Class. Quantum Gravity **19**, 897 (2002)
46. A. Gillespie, F. Raab, Thermally excited vibrations of the mirrors of laser interferometer gravitational-wave detectors. Phys. Rev. D **52**, 577 (1995)
47. Y. Levin, Internal thermal noise in the LIGO test masses: a direct approach. Phys. Rev. D **57**, 659 (1998)
48. V.B. Braginsky, M.L. Gorodetsky, S.P. Vyatchanin, Thermo-refractive noise in gravitational wave antennae. Phys. Rev. A **271**, 303 (2000)
49. T. Uchiyama, T. Tomaru, M.E. Tobar, D. Tatsumi, S. Miyoki, M. Ohashi, K. Kuroda, T. Suzuki, N. Sato, T. Haruyama, A. Yamamoto, T. Shintomi, Mechanical quality factor of a cryogenic sapphire test mass for gravitational wave detectors. Phys. Lett. A. **261**, 5 (1999)
50. A. Schroeter, R. Nawrodt, R. Schnabel, S. Reid, I. Martin, S. Rowan, C. Schwarz, T. Koettig, R. Neubert, M. Thürk, W. Vodel, A. Tünnermann, K. Danzmann, P. Seidel, *On the Mechanical Quality Factors of Cryogenic Test Masses from Fused Silica and Crystalline Quartz* (2007), arXiv:0709.4359v1
51. D. Friedrich, B.W. Barr, F. Brückner, S. Hild, J. Nelson, J. Macarthur, M.V. Plissi, M.P. Edgar, S.H. Huttner, B. Sorazu, S. Kroker, M. Britzger, E.-B. Kley, K. Danzmann, A. Tünnermann, K.A. Strain, R. Schnabel, Waveguide grating mirror in a fully suspended 10 meter Fabry-Perot cavity. Opt. Express **19**, 14955 (2011)
52. R. Abbott, R. Adhikari, G. Allen, S. Cowley, E. Daw, D. DeBra, J. Giaime, G. Hammond, M. Hammond, C. Hardham, J. How, W. Hua, W. Johnson, B. Lantz, K. Mason, R. Mittleman, J. Nichol, S. Richman, J. Rollins, D. Shoemaker, G. Stapfer, R. Stebbins, Seismic isolation for advanced LIGO. Class. Quantum Gravity **19**, 1591 (2002)
53. P.R. Saulson, Terrestrial gravitational noise on a gravitational wave antenna. Phys. Rev. D **30**, 732 (1984)
54. S.A. Hughes, K.S. Thorne, Seismic gravity-gradient noise in interferometric gravitational-wave detectors. Phys. Rev. D **58**, 122002 (1998)
55. V.B. Braginsky, M.L. Gorodetsky, S.P. Vyatchanin, Thermodynamical fluctuations and photo-thermal shot noise in gravitational wave antennae. Phys. Rev. A **264**, 1 (1999)
56. M. Zucker, S. Whitcomb, Measurement of optical path fluctuations due to residual gas in the LIGO 40 meter interferometer, in *Proceedings of Seventh Marcel Grossman Meeting on General Relativity*, (1996) pp. 1434–1436
57. E. Flanagan, K.S. Thorne, *Noise Due to Backscatter Off Baffles, the Nearby Wall, and Objects at the Far end of the Beam Tube; and Recommended Actions* (1994). LIGO-T940063-00-R
58. E. Flanagan, K.S. Thorne, *Scattered-Light Noise for LIGO* (1995). LIGO-T950132-00-R
59. T. Tomaru, Y. Saito, T. Kubo, Y. Sato, M. Tokunari, R. Takahashi, T. Suzuki, Y. Higashi, T. Shintomi, Y. Naito, N. Sato, T. Haruyama, A. Yamamoto, Study of optical dumpers used in high vacuum system of interferometric gravitational wave detectors. J. Phys. Conf. Ser. **32**, 476–481 (2006)

# Chapter 3
# Quantum Optics and Light

**Abstract** Quantum optics addresses the phenomena that can only be formalised by treating light as streams of discrete quanta of energy, or photons (Fox, Quantum Optics: An Introduction, 1996, [1]). Important for the work contained in this thesis, this chapter provides the background to quantum optics, and the modelling of common optical experiment components within that framework. This background material is based on the comprehensive text of Walls and Milburn (Quantum Optics, 2008, [2]). The basic quantum states of light (Sect. 3.2) and their pictorial representations (Sect. 3.3) are presented, followed by the quantum optical descriptions of mirrors and photodetection (Sect. 3.4), and optical cavities (Sect. 3.5).

## 3.1 Light Quantisation

### 3.1.1 The Quantised Electromagnetic Field

Derived from the source-free Maxwell's Equations, the quantised electromagnetic field (as a function of position **r** and time $t$) can be described by:

$$\mathbf{E}(\mathbf{r}, t) = i \sum_k \sqrt{\frac{\hbar\omega_k}{2\varepsilon_0}} [\hat{a}_k \mathbf{u_k}(\mathbf{r}) e^{-i\omega_k t} - \hat{a}_k^\dagger \mathbf{u_k^*}(\mathbf{r}) e^{i\omega_k t}] \tag{3.1}$$

where $\omega_k$ is the mode angular frequency, $\hbar$ is the reduced Planck constant, $\varepsilon_0$ the permittivity of free space, $\mathbf{u_k}(\mathbf{r})$ is the orthonormal spatial mode function set, and $\hat{a}_k$ and $\hat{a}_k^\dagger$ respectively the dimensionless boson annihilation and creation operators [2]. The annihilation and creation operators satisfy the boson commutation relations

$$[\hat{a}_k, \hat{a}_{k'}] = [\hat{a}_k^\dagger, \hat{a}_{k'}^\dagger] = 0, \qquad [\hat{a}_k, \hat{a}_{k'}^\dagger] = \delta_{kk'} \tag{3.2}$$

The quantised nature is supported by using Eq. 3.1 in writing the Hamiltonian $\hat{\mathscr{H}}$ for the total energy of the electromagnetic field. The results is



S.S.Y. Chua, *Quantum Enhancement of a 4 km Laser Interferometer Gravitational-Wave Detector*, Springer Theses,
DOI 10.1007/978-3-319-17686-4_3

$$\hat{\mathscr{H}} = \sum_k \hbar\omega_k \left(\hat{a}_k^\dagger \hat{a}_k + \frac{1}{2}\right) \tag{3.3}$$

The Hamiltonian is of the form of an ensemble of simple harmonic oscillators with quantised energy levels.

### 3.1.2 Quadrature Operators

The boson creation and annihilation operators are non-Hermitian, that is, they do not represent any observable or physically-measurable quantities. However, they can be re-defined as

$$\hat{a} = \frac{\hat{X}_1 + i\hat{X}_2}{2} \tag{3.4}$$

$$\hat{a}^\dagger = \frac{\hat{X}_1 - i\hat{X}_2}{2} \tag{3.5}$$

with the observable (thus Hermitian) non-commuting operator pair for the *amplitude quadrature*, $\hat{X}_1$, and *phase quadrature*, $\hat{X}_2$. Rearranging for the quadrature operators gives

$$\hat{X}_1 = \hat{a} + \hat{a}^\dagger \tag{3.6}$$

$$\hat{X}_2 = i(\hat{a} - \hat{a}^\dagger) \tag{3.7}$$

A linear combination of $\hat{X}_1$ and $\hat{X}_2$ can be used to describe an arbitrary quadrature $\hat{X}_\theta$,

$$\begin{aligned}\hat{X}_\theta &= \hat{a}e^{-i\theta} - \hat{a}^\dagger e^{i\theta} \\ &= \hat{X}_1 \cos\theta + \hat{X}_2 \sin\theta\end{aligned} \tag{3.8}$$

### 3.1.3 The Heisenberg Uncertainty Principle and Quantum Noise

From the statistical interpretation of quantum mechanics, the Heisenberg Uncertainty Principle defines the intrinsic uncertainty for two non-commuting observables, where precise information for both observables simultaneously is not possible [3]. Explicitly, for two observables that are related by the commutation relation

$$[\hat{O}_i, \hat{O}_j] = \delta_{ij} \tag{3.9}$$

their Uncertainty relation will be described by

$$\Delta\hat{O}_i\Delta\hat{O}_j \geq \frac{1}{2}|\delta| \tag{3.10}$$

where $\Delta\hat{O}$ is the standard deviation of the respective operator. The standard deviation is calculated from the Variance $V(\hat{O})$

$$V(\hat{O}) = (\Delta\hat{O})^2 = \sqrt{\langle\hat{O}^2\rangle - \langle\hat{O}\rangle^2} \tag{3.11}$$

The boson commutation relations (Eq. 3.2) translates the Uncertainty Principle to quantum optics. Using the creation and annihilation operators, this results in a commutation relation and uncertainty relation of

$$[\hat{a}, \hat{a}^\dagger] = 1 \tag{3.12}$$

$$\Delta\hat{a}\Delta\hat{a}^\dagger \geq \frac{1}{2} \tag{3.13}$$

and subsequently using the amplitude and phase quadratures, converts to

$$[\hat{X}_1, \hat{X}_2] = 2i \tag{3.14}$$

$$\Delta\hat{X}_1\Delta\hat{X}_2 \geq 1 \tag{3.15}$$

Therefore, it is not possible to precisely measure amplitude and phase quadratures simultaneously. It is this uncertainty that is referred to as the *quantum noise of the electromagnetic field*. If the fluctuations of a state satisfy the lower bound of the Uncertainty principle, namely $\Delta\hat{X}_1\Delta\hat{X}_2 = 1$, the state is said to be a *minimum uncertainty state*.

### 3.1.4 The Number Operator

An operator useful in describing states is the operator for the number of photons in a mode, or the *number operator*, defined from the creation and annihilation operators [1] by

$$\hat{N} = \hat{a}^\dagger\hat{a} \tag{3.16}$$

Calculating the mean photon number $\bar{N} = \langle\hat{N}\rangle$ gives, via multiplication with energy per photon $\hbar\omega$, the mean optical power within the mode.

$$P_{\text{opt}} = \bar{N}\hbar\omega \tag{3.17}$$

## 3.2 Important States of Light

The states of light of particular interest to quantum optic experiments are introduced here. These are the coherent state, the vacuum state, the squeezed state and the non-minimum uncertainty state. Diagrammatic representations of the various states are shown in Sect. 3.3.

### 3.2.1 The Coherent State

Light generated by a frequency and intensity stabilised laser source can be well approximated by a *coherent state* [2], thus making them important states of light for quantum optics experiments.

Coherent states, denoted by $|\alpha\rangle$, are both left eigenstates of the annihilation operator

$$\hat{a}|\alpha\rangle = \alpha|\alpha\rangle \tag{3.18}$$

and right eigenstates of the creation operator

$$\langle\alpha|\hat{a}^{\dagger} = (\hat{a}|\alpha\rangle)^{\dagger} = (\alpha|\alpha\rangle)^{\dagger} = \langle\alpha|\alpha^{*} \tag{3.19}$$

thus the mean photon number within a coherent state is

$$\begin{aligned}\bar{N} &= \langle\hat{N}\rangle \\ &= \langle\alpha|\hat{a}^{\dagger}\hat{a}|\alpha\rangle \\ &= \alpha^{*}\alpha\langle\alpha|\alpha\rangle \\ &= |\alpha|^{2}\end{aligned} \tag{3.20}$$

hence the term $\alpha$ is known as the *coherent amplitude* of the state [1].

Coherent states have fluctuations equally distributed across both quadratures and satisfy the lower bound of the Uncertainty principle, thus are minimum uncertainty states.

$$\Delta X_1 = \Delta X_2 = 1 \tag{3.21}$$

For measurements using unmodified coherent states, this minimum uncertainty represents the greatest accuracy in measurement that can be achieved. This limit is often described as the *quantum noise limit*.

### 3.2.2 The Vacuum State

Consider a special case of the coherent state that has no coherent amplitude ($\alpha = 0$). This state containing no photons on average is named a *vacuum state* and is denoted by $|0\rangle$. Despite having a mean photon number of zero, this does not result in the uncertainty in the amplitude and phase quadratures being zero, as that would violate the Heisenberg Uncertainty Principle. The uncertainties can be calculated for the vacuum state using Eq. 3.11, that is, for the amplitude quadrature variance

$$\begin{aligned} V(\hat{X}_1|0\rangle) &= \langle 0|(\hat{X}_1)^2|0\rangle - (\langle 0|\hat{X}_1|0\rangle)^2 \\ &= \langle 0|(\hat{a}+\hat{a}^\dagger)^2|0\rangle - (\langle 0|\hat{a}+\hat{a}^\dagger|0\rangle)^2 \\ &= \langle 0|(\hat{a}^2+\hat{a}^\dagger\hat{a}+\hat{a}\hat{a}^\dagger+\hat{a}^{\dagger 2})|0\rangle - 0 \\ &= 0+0+\langle 0|(1+\hat{a}^\dagger\hat{a})|0\rangle + 0 \\ &= 1+0 \\ &= 1 \\ \Delta X_1 &= \pm 1 \end{aligned} \tag{3.22}$$

and similarly for the phase quadrature

$$\Delta X_2 = \pm 1 \tag{3.23}$$

The above calculation shows that the vacuum state is a minimum uncertainty state. This state of light is present in any region in space where there is no occupying bright light state. These fluctuations manifest in the energy Hamiltonian $\hat{\mathscr{H}}$ (Eq. 3.3) via the $\frac{1}{2}\hbar\omega_k$ term per oscillator.

### 3.2.3 The Squeezed State

Consider two generic operators $\hat{J}$ and $\hat{K}$ that follows the commutator relation $[\hat{J}, \hat{K}] = i\hat{L}$. From the Heisenberg Uncertainty Principle, it follows that

$$\langle \Delta \hat{J}^2\rangle\langle \Delta \hat{K}^2\rangle \geq \frac{1}{4}|\langle \hat{L}\rangle|^2 \tag{3.24}$$

The Uncertainty relation is a multiplicative limit, only constraining the product of the variances. However, it does not constrain the individual component variances, thus if

$$\langle \Delta \hat{J}\rangle^2 < \frac{1}{2}|\langle \hat{L}\rangle| \quad \text{or} \quad \langle \Delta \hat{K}\rangle^2 < \frac{1}{2}|\langle \hat{L}\rangle| \tag{3.25}$$

but not simultaneously (else Eq. 3.24 is violated), the state is then said to be *squeezed* [2]. Thus a squeezed state refers to a state where the uncertainty of one quadrature is below the lower bound of the respective Uncertainty relation. Correspondingly the standard deviation in the other quadrature will be greater than the lower bound.

More formally, the unitary squeeze operator is defined as [2]

$$\hat{S}(\zeta) = \exp\left(1/2(\zeta^* \hat{a}^2 - \zeta \hat{a}^{\dagger 2})\right) \tag{3.26}$$

where $\zeta = Re^{2i\Psi}$. The degree of squeezing is set by the squeeze factor $R = |\zeta|$ while the quadrature angle of the squeezing is set by $\Psi$. For an arbitrary quadrature $\hat{X}_\theta$, the quadrature uncertainties can be described by

$$\langle \Delta \hat{X}_\theta \rangle = e^{-R} \tag{3.27}$$

$$\langle \Delta \hat{X}_{\theta+\pi/2} \rangle = e^{+R} \tag{3.28}$$

which satisfies the minimum uncertainty relation.

For squeezed states, the mean photon number is described by

$$\bar{N} = |\alpha|^2 + \sinh^2 R \tag{3.29}$$

Both coherent states and vacuum states can be squeezed, named *bright squeezed* and *squeezed vacuum* states respectively. Squeezed vacuum states are the focus of this thesis.

The implication of Eq. 3.29 is that although $\alpha = 0$, for squeezed vacuum states the mean photon number is non-zero, thus there is optical power in the squeezed vacuum beam. The amount of power is dependent on the magnitude of the squeezing (from the squeezed factor $R$). This optical power can be viewed as the energy necessary to change the noise statistics of the vacuum state. When considering the Quantum Sideband Diagram of squeezed vacuum states (Sect. 3.3.3), accounting for this optical power will be necessary.

### 3.2.4 Non-minimum Uncertainty States

Non-minimum uncertainty states have fluctuations that are greater than the lower bound product of the Uncertainty Principle. The noise is often in both quadratures, and is often greater than the quantum noise limit, that is

$$\Delta X_1 > 1, \quad \Delta X_2 > 1 \tag{3.30}$$

Examples of such states include squeezed states mixed with additional noise, and lasers with excess classical noise. Noise suppression techniques are used (for example with lasers [4, 5]) to return the states closer to the quantum noise limit.

## 3.3 Pictorial Representations of the States of Light

After an overview of their classical counter-part diagrams, this section introduces the common pictorial representations of quantum states of light—the quantum phasor diagram and the quantum sideband diagram.

### 3.3.1 Classical Phasor and Classical Sideband Diagrams

We begin by taking a classical electric field at a given time, described by

$$E(t) = E_0 e^{i\omega_0 t} \tag{3.31}$$

This field can be illustrated on the complex plane as a vector of magnitude $E_0$, with the coordinate system rotating in-step with vector phase $\omega_0 t$, known as the *rotating frame (of reference)*. This picture is called the *classical phasor diagram*, as shown in Fig. 3.1a. Should there be multiple vectors, one vector is chosen to be the stationary reference vector of the rotating frame, while other vectors vary in phase with respect to that reference. Typically, the chosen reference vector is the *carrier* field [6].

In optics, the application of *modulation* describes the process by which a carrier field is modulated by an external time-varying waveform, via a mechanical or optical actuator [6]. First consider a field with *amplitude modulation* at modulation frequency $\omega_m$ and modulation depth $M$, described by [7]

$$\begin{aligned} E_{\text{a-mod}}(t) &= E_0 e^{i\omega_0 t}(1 + M\cos(\omega_m t)) \\ &= E_0 e^{i\omega_0 t}(1 + \frac{M}{2}e^{i\omega_m t} + \frac{M}{2}e^{-i\omega_m t}) \end{aligned} \tag{3.32}$$

From Eq. 3.32, the amplitude modulated field is decomposed into a carrier field and two *sidebands* at frequencies $\omega_0 \pm \omega_m$. This is shown pictorially in Fig. 3.1b1, where the chosen reference is the rotating frame of the carrier, and the sidebands rotating in opposite directions around the carrier. We can also apply the same methodology to a field with *phase modulation*, described by

$$E_{\text{p-mod}}(t) = E_0 e^{i(\omega_0 t + M\cos(\omega_m t))})$$

$$\approx E_0 e^{i\omega_0 t}(1 + iM\cos(\omega_m t)) \tag{3.33}$$

$$= E_0 e^{i\omega_0 t}(1 + i\frac{M}{2}e^{i\omega_m t} + i\frac{M}{2}e^{-i\omega_m t}) \tag{3.34}$$

and shown in Fig. 3.1b2. The approximation in Eq. 3.33 is valid for small values of the modulation depth—the complete description utilises Bessel Functions of the First Kind ($J_n$) [7]. The pictorial representation of a field with a carrier and sidebands is called the *classical sideband diagram*. The links between classical sideband

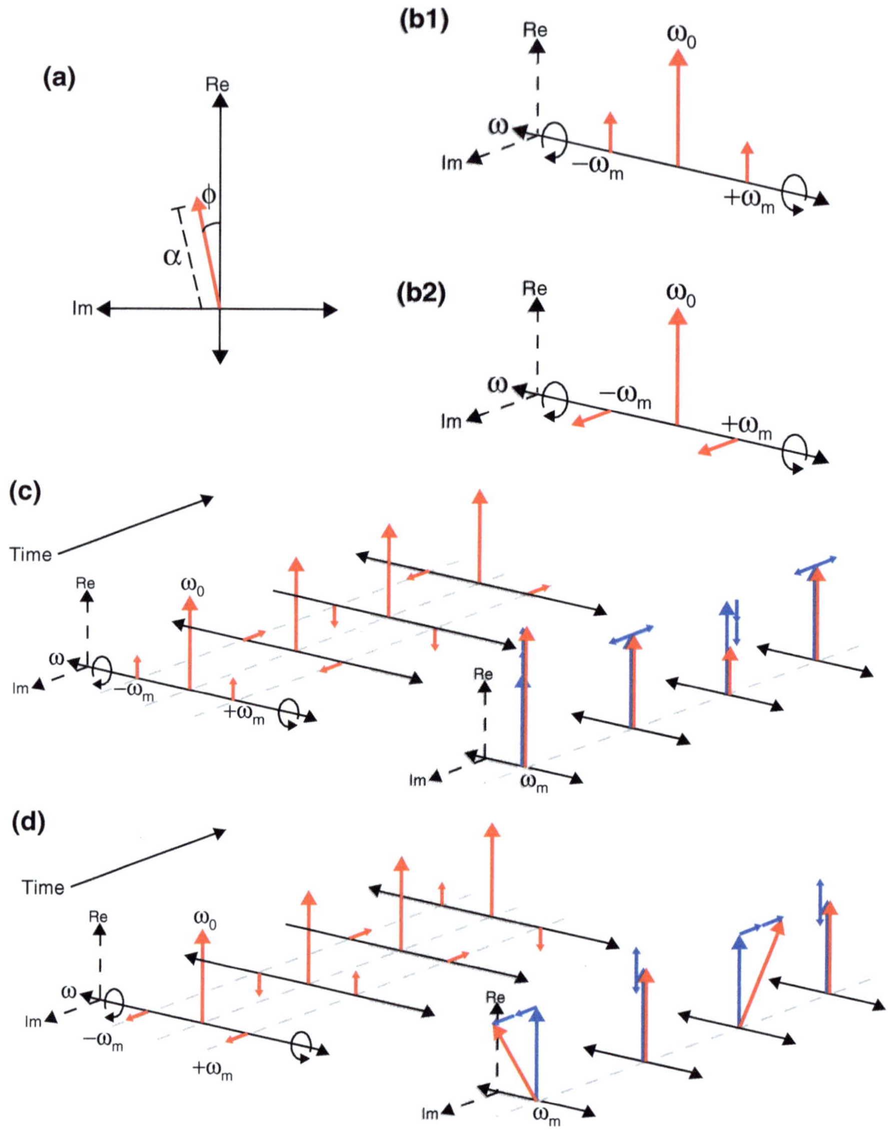


**Fig. 3.1** **a** Classical phasor diagram with phasor with amplitude $\alpha$ and phase $\phi$, **b** classical sideband diagram for amplitude and phase modulation described by Eqs. 3.32 and 3.34 respectively, **c** time evolution of amplitude modulation sidebands and resulting phasor evolution (changing $\alpha$), and **d** time evolution of phase modulation sidebands and resulting phasor evolution (changing $\phi$). The modulation effect is exaggerated for clarity

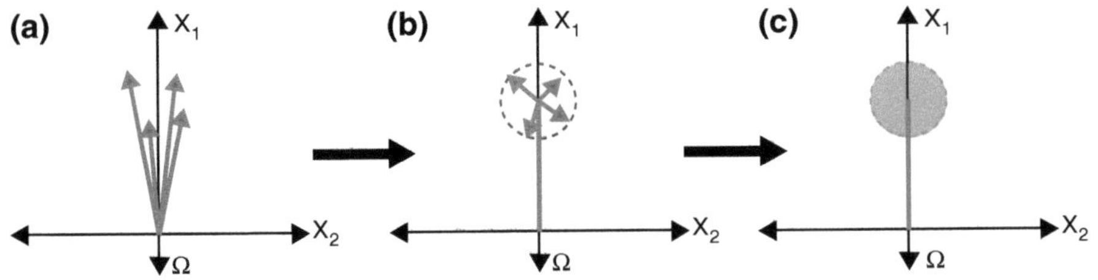


**Fig. 3.2** Quantum phasor diagram of a coherent state. **a** Multiple phasors representing different possible amplitudes and phases, **b** equivalent single phasor with time-varying, non-deterministic uncertainty vector, and **c** the quantum phasor ("Ball on Stick") diagram with phasor "stick" and quadrature uncertainty "ball". Reprinted with permission from [8], copyright IOP Publishing (2014)

and classical phasor are demonstrated with the time-evolution series of diagrams of Fig. 3.1c, d (amplitude and phase modulation respectively) and how the sum of the carrier and sidebands changes the classical phasor.

### 3.3.2 Quantum Phasor Diagram

From the Heisenberg Uncertainty Principle, there is uncertainty in the amplitude and phase quadratures (Sect. 3.1.3), thus with a coherent state as an example, at a measurement frequency of $\Omega$, there are now many different phasors of varying amplitudes and phases possible, as shown in Fig. 3.2a. These many phasors can be equivalently described by a single phasor with an added non-deterministic, time-varying amplitude/phase uncertainty vector, as shown in Fig. 3.2b. As the uncertainty in both quadratures are equal ($\Delta X_1 = \Delta X_2 = 1$), we can illustrate the uncertainty vector with a two-dimensional gaussian distribution, thus realising the *quantum phasor diagram* shown in Fig. 3.2c. This diagram is a hybrid diagram in quadrature space ($X_1$, $X_2$) that illustrates both classical and quantum properties of the state. Quantum phasor diagrams of various states are shown in Fig. 3.3. The states with squeezed quadratures are distinctive with an elliptical distribution.

### 3.3.3 Quantum Sideband Diagram

Consider the progression illustrated in Fig. 3.4a. We begin by returning to the quantum phasor description of the form of a single phasor and an added uncertainty vector (Fig. 3.2b). Utilising the classical sideband picture, we can represent the added vector as quantum sidebands at frequency $\omega_0 \pm \Omega$, and the amplitude/phase uncertainty is encompassed by the time-varying *orientation* of these sidebands with respect to each other. Expanding to include all frequencies $\omega_0 \pm \Omega,\ \Omega \in \mathbb{R}^+$, this completes the

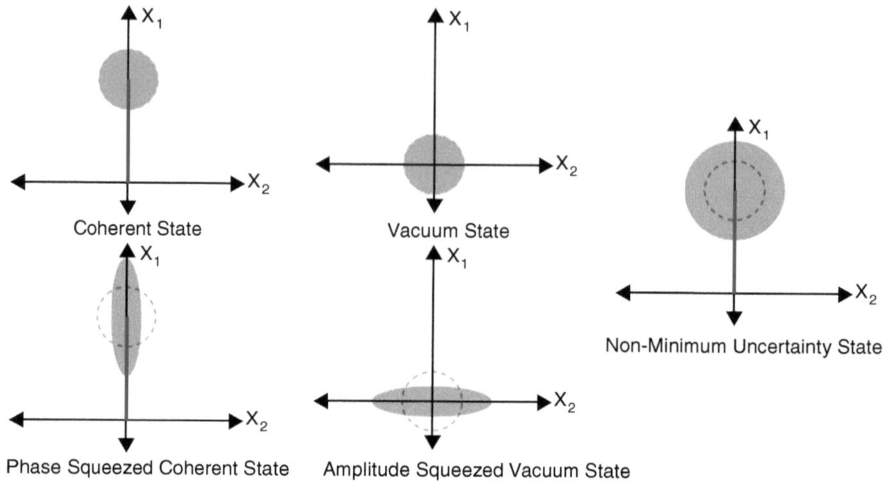


**Fig. 3.3** Quantum phasor diagram representations of various quantum light states

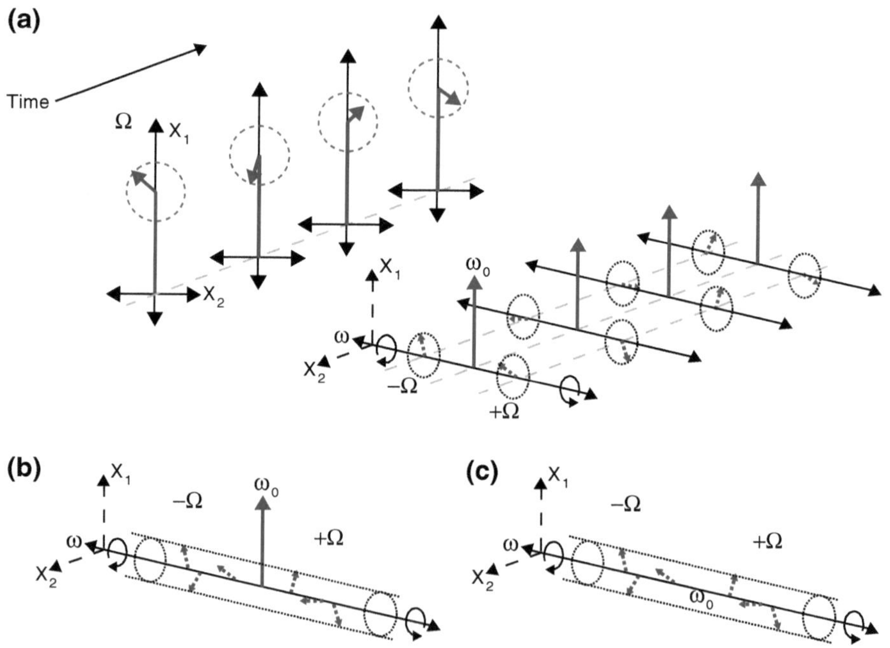


**Fig. 3.4** **a** Progression from the Quantum Phasor Diagram to the Quantum Sideband Diagram. Quantum Sideband Diagrams for the **b** Coherent State and **c** Vacuum State

*quantum sideband diagram*. These sidebands are a visualisation of the $\frac{1}{2}\hbar\omega_k$ energy terms of the Hamiltonian (Eq. 3.3). The quantum sideband diagrams of the coherent state and vacuum state are shown in Fig. 3.4b, c respectively.

From Eq. 3.29, a squeezed vacuum state contains photons. These photons replace the (unsqueezed) vacuum state, and this can be visualised with *correlated sideband pairs*. The magnitude of squeezing determines how well these sideband pairs are correlated. In the quantum sideband diagram, we can represent squeezing with perfectly correlated sideband phasors with two-dimensional gaussian distributions representing the degree of correlation—the smaller the distribution, the stronger the correlation. Phase and Amplitude vacuum squeezed states, in the quantum sideband diagram leading to the quantum phasor diagram, are depicted in Fig. 3.5.

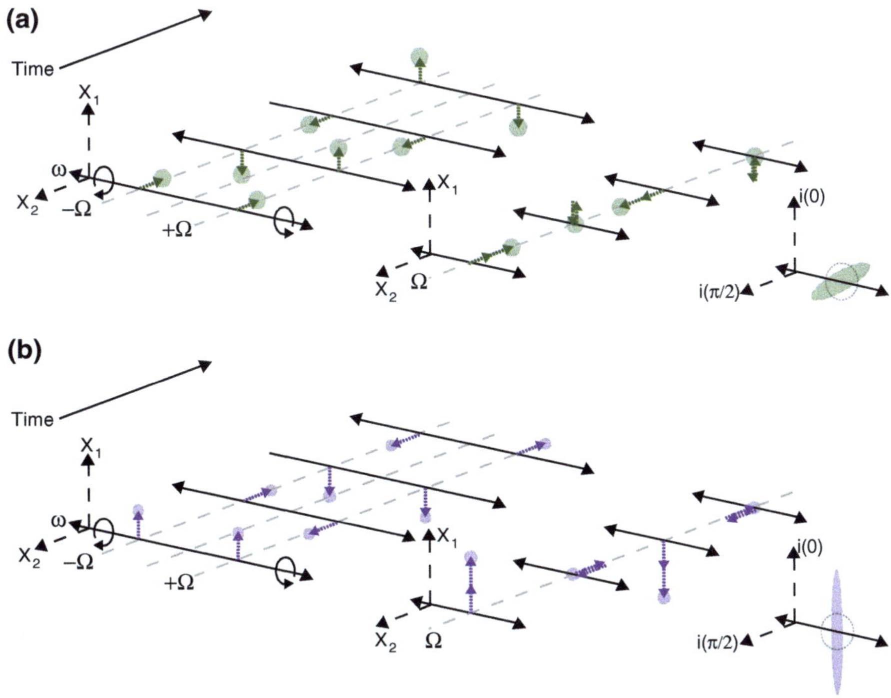


**Fig. 3.5** Time evolution of correlated sidebands at frequency $\omega_0 \pm \Omega$ in the quantum sideband picture, and resulting measured (quantum) phasor, for an example **a** amplitude squeezed vacuum state and **b** phase squeezed vacuum state. The measurement quadrature $i(0)$ is aligned with the amplitude quadrature $X_1$. The smaller uncertainty distribution, combined with larger sidebands due to greater power (Eq. 3.29) in (**b**) indicate a stronger magnitude of squeezing. The result is a measured phasor with a more elongated squeezing noise ellipse

## 3.4 A Quantum Treatment of Optical Components

### 3.4.1 Linearisation of Operators

A commonly used method to obtain analytic results in quantum optics is to apply a linearisation procedure to operators. This was first applied by Yurke in 1984 [9].

When linearised, an operator $\hat{O}$ is now separable into two components and is described as follows:

$$\hat{O}(t) = \bar{O} + \delta\hat{O}(t), \quad \hat{O}^\dagger(t) = \bar{O}^* + \delta\hat{O}^\dagger(t) \tag{3.35}$$

where $\bar{O} = \langle\hat{O}\rangle$, $\bar{O}^* = \langle\hat{O}^\dagger\rangle$ are the steady-state components and $\delta\hat{O}(t) = \bar{O} - \hat{O}(t)$, $\delta\hat{O}^\dagger(t) = \bar{O}^* - \hat{O}^\dagger(t)$ are the fluctuating components. It is assumed that on average, the fluctuating components have no constant amplitude and have a magnitude much smaller than their steady-state counterparts, namely:

$$\langle\delta\hat{O}(t)\rangle = \langle\delta\hat{O}^\dagger(t)\rangle = 0 \tag{3.36}$$

$$|\delta\hat{O}(t)| \ll \bar{O} \qquad |\delta\hat{O}^\dagger(t)| \ll \bar{O}^* \tag{3.37}$$

The first assumption allows for no net contribution to the constant amplitude component from the fluctuations. The second assumption allows the simplification of making a first order approximation, as higher order terms now have negligible contributions. The commutation relations thus no longer have any bearing and a complete semiclassical, analytical expression of a quantum operator/variable is accessible [2].

Applying the linearisation formalism to the creation and annihilation operators, we obtain the linearised form of the fluctuations in the phase and amplitude quadratures. These are

$$\hat{a}(t) = \bar{a} + \delta\hat{a}(t) \tag{3.38}$$

$$\hat{a}^\dagger(t) = \bar{a}^* + \delta\hat{a}^\dagger(t) \tag{3.39}$$

$$\delta\hat{X}_1(t) = \delta\hat{a}(t) + \delta\hat{a}^\dagger(t) \tag{3.40}$$

$$\delta\hat{X}_2(t) = i(\delta\hat{a}(t) - \delta\hat{a}^\dagger(t)) \tag{3.41}$$

### 3.4.2 A Partially Transmissive Mirror

Figure 3.6a shows two generic fields A and B incident on a lossless partially-transmissive mirror, or lossless *beamsplitter*, with transmission $\varepsilon$. The fields at the outputs of the mirror, C and D, can be related to the input fields via

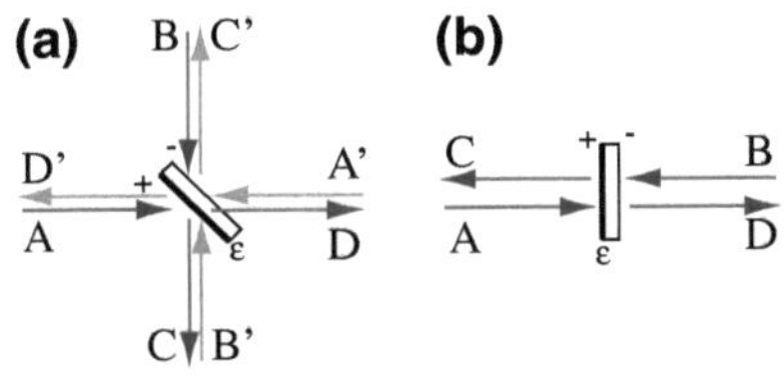


**Fig. 3.6** Partially transmissive mirrors. The fields $A, B, C, D$ are labelled for a partially transmissive mirror with fields at **a** 45° incidence and **b** at normal incidence, drawn spatially separated for clarity. The + and − signs indicates the side where the amplitude reflectivity receives a minus sign

$$C = \sqrt{1-\varepsilon}A + \sqrt{\varepsilon}B \tag{3.42}$$
$$D = \sqrt{\varepsilon}A - \sqrt{1-\varepsilon}B \tag{3.43}$$

where the convention used (here and for this thesis) to maintain the phase relations of the beamsplitter ports is to set the reflectivity of the mirror to be $-1 \times \sqrt{1-\varepsilon}$ on one side. This is indicated by a minus sign on a diagram. The other commonly used convention to maintain the phase relations of the beamsplitter ports is to multiply a transmitted field by $i = \sqrt{-1}$ on each transmission pass.

A second set of input and output beams are also present, indicated by A′, B′, C′ and D′ in Fig. 3.6a. With normal incidence as shown in Fig. 3.6b, the two sets of inputs/outputs become a single set of inputs/outputs as fields become equivalent, for example field A is equivalent to field B′.

### 3.4.3 Photodetection

Direct detection of a light beam of interest (using a photodiode) is a very common technique used in optical experiments. A basic illustration is shown in part (a) of Fig. 3.7.

As a consequence of the Photoelectric Effect, photons incident onto a photosensitive medium produce a source of freely moving electrons [6]. The produced *photocurrent* $i$, is given by

$$i = \rho P_{\text{opt}} = \frac{e\,\eta_{pd}}{\hbar\omega} P_{\text{opt}} \tag{3.44}$$

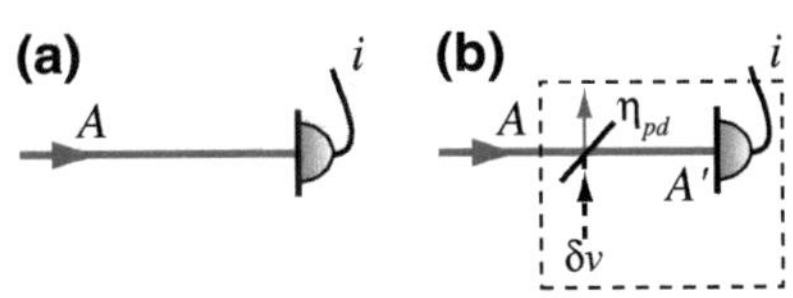


**Fig. 3.7** Photodetection: **a** Light field $A$ incident onto a perfect, noiseless photodetector. **b** A real-world photodetector, with quantum detection efficiency $\eta_{pd}$

where $\rho$ is the *responsivity* of the photodetector (in units of A/W), $\eta_{pd}$ is the *quantum detection efficiency* of the detection medium, $\hbar\omega$ the energy per photon, and $e$ is the modulus of the electron charge.

The photocurrent generated by the photodiode is proportional to the number of photons in the optical field, thus using the number operator (Sect. 3.1.4) and combined with the linearisation formalism (Sect. 3.4.1):

$$\begin{aligned} N_A &= \langle A|\hat{a}^\dagger \hat{a}|A\rangle \\ &= \langle A|(\langle \hat{a}^\dagger\rangle + \delta\hat{a}^\dagger)(\langle \hat{a}\rangle + \delta\hat{a})|A\rangle \\ &= (A^* + \delta A^\dagger)(A + \delta A) \\ &= |A|^2 + A(\delta A^\dagger + \delta A) \\ &= |A|^2 + A(\delta X_1^A) \end{aligned} \tag{3.45}$$

Thus for a detector with perfect efficiency $\eta_{pd} = 1$, the photocurrent is

$$\begin{aligned} i &= \frac{e}{\hbar\omega} P_{\text{opt}} \\ &= \frac{e}{\hbar\omega} N\hbar\omega \\ &= e(|A|^2 + A(\delta X_1^A)) \end{aligned} \tag{3.46}$$

From left to right, the above expression consists of a non-fluctuating term that is directly proportional to the optical intensity of the incident light, and a fluctuating term that is proportional to the amplitude quadrature times the coherent amplitude i.e.:

$$\langle i\rangle \propto |A|^2, \qquad \delta i \propto A(\delta X_1^A) \tag{3.47}$$

The analysis has proceeded thus far assuming perfect photodetection. In physical experiments however, noise and inefficiencies must be taken into account. The model of physical photodetectors involves placing an imaginary beamsplitter with transmittivity equal to $\eta_{pd}$ before the perfect photodetector, as shown in part (b) of Fig. 3.7. The photocurrent becomes:

$$\begin{aligned} i = e\langle A'|\hat{a}^\dagger\hat{a}|A'\rangle &= e\left(\langle \sqrt{\eta_{pd}}A^\dagger + \sqrt{1-\eta_{pd}}\delta v^\dagger|\hat{a}^\dagger\hat{a}|\sqrt{\eta_{pd}}A + \sqrt{1-\eta_{pd}}\delta v\rangle\right) \\ &= e\left(\eta_{pd}(|A|^2 + A(\delta a^\dagger + \delta a)) + \sqrt{\eta_{pd}(1-\eta_{pd})}A(\delta v^\dagger + \delta v)\right) \\ &= e\left(\eta_{pd}(|A|^2 + A\delta X_1^a) + \sqrt{\eta_{pd}(1-\eta_{pd})}A\delta X_1^v\right) \end{aligned} \tag{3.48}$$

Therefore, the photocurrent components are

$$\langle i\rangle = e\eta_{pd}|A|^2, \qquad \delta i = eA\left(\eta_{pd}\delta X_1^a + \sqrt{\eta_{pd}(1-\eta_{pd})}\delta X_1^v\right) \tag{3.49}$$

The two effects of non-perfect detection are now evident. The first effect is that only transmitted photons are counted, factored in by the transmission of the imaginary beamsplitter. The second effect is that optical losses are accompanied by fluctuations that degrade the quantum properties of the detected light, accounted for by the entrance of vacuum fluctuations $\delta v$ into the second input port of the imaginary beamsplitter.

## 3.5 Optical Cavities

An *optical cavity* is an arrangement of partially transmissive mirrors that allow an optical beam to circulate in a closed path. In this section the quantum Langevin formalism that is used to describe the behaviour of optical cavities is introduced. Key cavity parameters, and the cavity response in reflection and transmission are also discussed.

### 3.5.1 Quantum Langevin Equations of an Optical Cavity

A generic optical cavity is shown in Fig. 3.8. The external cavity fields are denoted with capitalised notation, i.e. $A$, with units $\sqrt{\text{photons}/s}$. The *circulating field* or *cavity mode* is denoted with lower-case notation, i.e. $a$, with scalar (photon) units. The three mirrors are labelled as an input-coupling mirror 'in' (or *input coupler*), an output-coupling mirror 'out' (or *output coupler*), and the third mirror utilised to separately account for any optical loss $l$ on the circulating field.

The cavity round trip time is defined by:

$$\tau = \frac{OPL}{c} \tag{3.50}$$

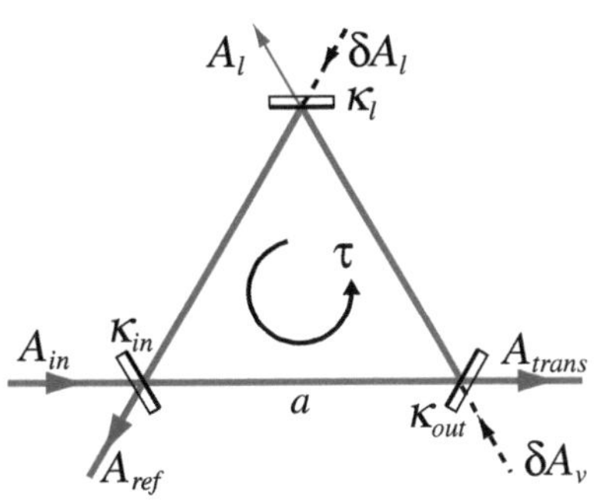


**Fig. 3.8** A generic optical cavity with the quantum Langevin formalism, described in Sect. 3.5.1

where *OPL* denotes the *round trip optical path length* and $c$ the speed of light. For each mirror $j(= in, out, l)$, the power reflectivity $\mathrm{R}_j$ and transmissivity $\mathrm{T}_j$ are defined by

$$\mathrm{R}_j = r_j^2, \qquad \mathrm{T}_j = t_j^2 = 1 - \mathrm{R}_j \tag{3.51}$$

The *finesse* of an optical cavity, denoted by $\mathscr{F}$, is a dimensionless quantity that describes the "narrowness" of the cavity resonance peaks. For an optical cavity with $i$ mirrors ($i \in \mathscr{N} \geq 2$), the finesse is given by

$$\mathscr{F} = \frac{\pi(R_1 R_2 \ldots R_i)^{\frac{1}{i+1}}}{1 - \sqrt{R_1 R_2 \ldots R_i}} \tag{3.52}$$

where $R_i$ is the power reflectivity of mirror $i$.
The *coupling rate* $\kappa_j$ for each mirror is defined from the transmissivity and round trip time, namely:

$$\kappa_j = \frac{t_j}{\tau} = \frac{\sqrt{T_j}}{\tau} \tag{3.53}$$

The total decay rate $\kappa$ of the cavity is the sum of the individual decay rates

$$\kappa = \sum \kappa_j \tag{3.54}$$

Optical fields external to the optical cavity are denoted with upper case letters, while internal or *circulating* cavity fields are denoted with lower case letters. Following on from Sect. 3.1.4, the optical power of an external cavity field $A_Z$ with angular frequency $\omega_Z$ is given by:

$$P_Z = \hbar\omega_Z |A_Z|^2 \tag{3.55}$$

and the circulating power of generic field $a_z$ within the cavity is given by:

$$P_z = \frac{\hbar\omega_z |z|^2}{\tau} \tag{3.56}$$

with the factor of the cavity round trip time ($\tau$) making $P_z$ unit-consistent (photons/s).

Using the formalism developed by Gardiner and Collett [10], the quantum Langevin equation of motion for the cavity mode, using the above defined parameters, is given by:

$$\dot{a} = -(\kappa + i\omega_0)a + \sqrt{2\kappa_{in}} A_{in} e^{-i\omega_A t} + \sqrt{2\kappa_{out}}\delta A_v + \sqrt{2\kappa_l}\delta A_l \tag{3.57}$$

with $A_{in}$ denoting the coherent field at frequency $\omega_A$, $\omega_0$ the cavity resonant frequency, and the other input fields being vacuum states $\delta A_v$ and $\delta A_l$. Equation 3.57 can be rewritten in the rotating frame of $\omega_A$, becoming:

$$\dot{a} = -(\kappa + i\Delta)a + \sqrt{2\kappa_{in}}A_{in} + \sqrt{2\kappa_{out}}\delta A_v + \sqrt{2\kappa_l}\delta A_l \tag{3.58}$$

with *cavity detuning* $\Delta = \omega_0 - \omega_A$, the difference between coherent field and cavity resonant frequency. Setting the cavity detuning to zero (i.e. $\Delta = 0$), which is now experimentally readily achievable, the Fourier transform of Eq. 3.58 gives:

$$i\omega\tilde{a} = -\kappa\tilde{a} + \sqrt{2\kappa_{in}}\tilde{A}_{in} + \sqrt{2\kappa_{out}}\delta\tilde{A}_v + \sqrt{2\kappa_l}\delta\tilde{A}_l \tag{3.59}$$

$$\Rightarrow \tilde{a} = \frac{\sqrt{2\kappa_{in}}\tilde{A}_{in} + \sqrt{2\kappa_{out}}\delta\tilde{A}_v + \sqrt{2\kappa_l}\delta\tilde{A}_l}{i\omega + \kappa} \tag{3.60}$$

where $\omega$ is a frequency shift about the optical carrier frequency. To calculate the reflected field ($A_{ref}$) and transmitted field ($A_{trans}$), the input-output relations [10] are used, given by

$$\tilde{A}_{ref} = \sqrt{2\kappa_{in}}\tilde{a} - \tilde{A}_{in} \tag{3.61}$$

$$\tilde{A}_{trans} = \sqrt{2\kappa_{out}}\tilde{a} - \delta\tilde{A}_v \tag{3.62}$$

which via substitution of $\tilde{a}$ results in:

$$\tilde{A}_{ref} = \frac{(2\kappa_{in} - \kappa - i\omega)\tilde{A}_{in} + \sqrt{2\kappa_{in}\kappa_{out}}\delta\tilde{A}_v + \sqrt{2\kappa_{in}\kappa_l}\delta\tilde{A}_l}{i\omega + \kappa} \tag{3.63}$$

$$\tilde{A}_{trans} = \frac{\sqrt{2\kappa_{in}\kappa_{out}}\tilde{A}_{in} + (2\kappa_{out} - \kappa - i\omega)\delta\tilde{A}_v + \sqrt{2\kappa_{out}\kappa_l}\delta\tilde{A}_l}{i\omega + \kappa} \tag{3.64}$$

We can now consider the steady-state and fluctuating outputs of the cavity by applying the linearisation formalism.

### 3.5.2 Reflected and Transmitted Fields

From Eqs. 3.63 and 3.64, the steady-state transfer functions in reflection and transmission are

$$\mathscr{R}(\omega) = \frac{\bar{\tilde{A}}_{ref}}{\bar{\tilde{A}}_{in}} = \frac{(2\kappa_{in} - \kappa - i\omega)}{i\omega + \kappa} \tag{3.65}$$

$$\mathscr{T}(\omega) = \frac{\bar{\tilde{A}}_{trans}}{\bar{\tilde{A}}_{in}} = \frac{\sqrt{2\kappa_{in}\kappa_{out}}}{i\omega + \kappa} \tag{3.66}$$

The transfer function magnitude response ($|\mathscr{R}(\omega)|^2$, $|\mathscr{T}(\omega)|^2$) and phase response ($\angle\mathscr{R}(\omega)$, $\angle\mathscr{T}(\omega)$) are shown in Fig. 3.9. This shows the response of a cavity to an

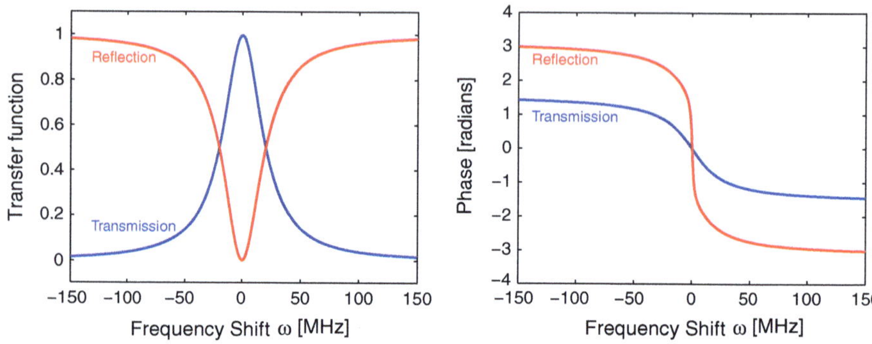


**Fig. 3.9** The magnitude and phase transfer functions of an optical cavity in reflection and transmission. Cavity parameters chosen are $L = 1$ m, $\lambda = 1$ μm, $T_{in} = 0.05$, $T_{out} = 0.04$, $T_l = 0$

incoming field is strongly dependent on the frequency of the incoming field relative to the resonant frequency of the cavity.

### *3.5.3 Noise Variances of the Reflected and Transmitted Fields*

Returning to Eqs. 3.63 and 3.64, the fluctuating components in reflection and transmission are

$$\delta\tilde{A}_{ref} = \frac{(2\kappa_{in} - \kappa - i\omega)\delta\tilde{A}_{in} + \sqrt{2\kappa_{in}\kappa_{out}}\delta\tilde{A}_v + \sqrt{2\kappa_{in}\kappa_l}\delta\tilde{A}_l}{i\omega + \kappa} \tag{3.67}$$

$$\delta\tilde{A}_{trans} = \frac{\sqrt{2\kappa_{in}\kappa_{out}}\delta\tilde{A}_{in} + (2\kappa_{out} - \kappa - i\omega)\delta\tilde{A}_v + \sqrt{2\kappa_{out}\kappa_l}\delta\tilde{A}_l}{i\omega + \kappa} \tag{3.68}$$

Constructing the quadratures ($\delta X_{1,2}$), then calculating the variances ($V_{1,2}$) gives

$$V_{1,2}^{ref} = \frac{((2\kappa_{in} - \kappa)^2 + \omega^2)V_{1,2}^{in} + 4\kappa_{in}\kappa_{out}V_{1,2}^{v} + 4\kappa_{in}\kappa_l V_{1,2}^{l}}{\omega^2 + \kappa^2} \tag{3.69}$$

$$V_{1,2}^{trans} = \frac{4\kappa_{in}\kappa_{out}V_{1,2}^{in} + ((2\kappa_{out} - \kappa)^2 + \omega^2)V_{1,2}^{v} + 4\kappa_{out}\kappa_l V_{1,2}^{l}}{\omega^2 + \kappa^2} \tag{3.70}$$

The above variances can be simplified by noting that for the transmission and loss ports, the entering flucutations are vacuum fluctuations, thus $V_{1,2}^{v} = V_{1,2}^{l} = 1$. In Fig. 3.10, the reflection and transmission noise response is plotted for an input amplitude variance of ten times the minimum uncertainty noise level, that is, $V_1^{in} = 10$. This shows that an optical cavity can act as a noise *cleaner*, a filter for noise. The noise response is of a low pass filter in transmission and a high pass filter in reflection.

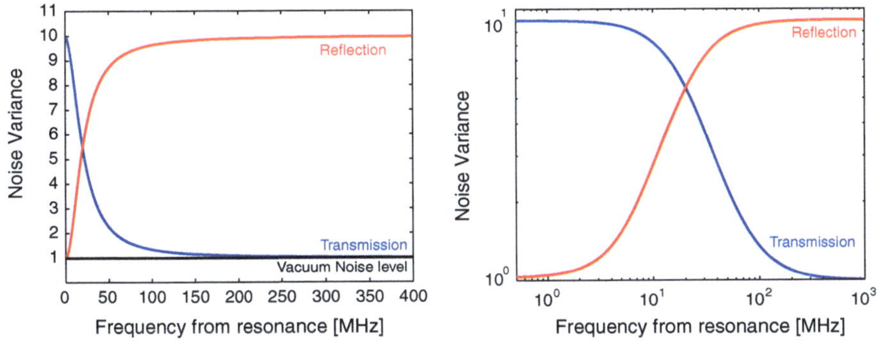


**Fig. 3.10** Example noise response of an optical cavity in reflection and transmission. The cavity parameters are the same as those used in Fig. 3.9

## 3.6 Summary

This chapter provides a background to the framework of quantum optics formalism. The basics of quantum states of light and their pictorial representations were reviewed. The quantum optical descriptions of mirrors, optical cavities and photodetection were also covered. These concepts will be utilised throughout the remainder of this thesis.

## References

1. M. Fox. *Quantum Optics: An Introduction* (Oxford University Press, Oxford, 1996)
2. D.F. Walls, G. Milburn, *Quantum Optics*, 2nd edn. (Springer, Berlin, 2008)
3. D.J. Griffiths, *Introduction to Quantum Mechanics*, 1st edn. (Prentice Hall, New Jersey, 1994)
4. S.A. Webster, M. Oxborrow, P. Gill, Subhertz-linewidth Nd:YAG laser. Opt. Lett. **29**(13), 1497–1499 (2004)
5. P. Kwee, B. Willke, K. Danzmann, Shot-noise-limited laser power stabilization with a high-power photodiode array. Opt. Lett. **34**(19), 2912–2914 (2009)
6. B.E.A. Saleh, M.C. Teich. *Fundamentals of Photonics* (Wiley, New York, 1991)
7. J.D. Gibson. *Principles of Digital and Analog Communications*, 2nd edn. (Prentice Hall, New Jersey, 1993)
8. S.S.Y. Chua, B.J.J. Slagmolen, D.A. Shaddock, D.E. McClelland, Quantum squeezed light in gravitationalwave detectors. Class. Quantum Grav. **31**, 183001 (2014)
9. B. Yurke, Use of cavities in squeezed state generation. Phys. Rev. A **20**, 408 (1984)
10. C.W. Gardiner, M.J. Collett, Input and output in damped quantum systems: quantum stochastic differential equations and the master equation. Phys. Rev. A **31**, 3761–3774 (1985)

# Chapter 4
# Quantum Noise in Gravitational-Wave Detectors and Applied Squeezed States

**Abstract** The development of low-frequency squeezed states is motivated by gravitational-wave detectors being limited by quantum noise. This chapter presents a background to the quantum noise limits in these detectors, along with the enhancements possible with squeezed state injection. The calculations presented are heavily based on the derivations from the works of Kimble et al. (Phys Rev D 65:022002, 2002, [1]) and Buonanno and Chen (Phys Rev D 64:042006, 2001; Phys Rev D 69:102004, 2004, [2, 3]). Cited frequently throughout this chapter, these works use estimated experiment parameters for LIGO and Advanced LIGO detectors. Section 4.1 provides an introduction of the influence of quantum noise in measurement, followed by a brief discussion in the main source of quantum noise in gravitational-wave detection (Sect. 4.2). Subsequently, the quantum noise of three Michelson configurations are explored: a simple Michelson interferometer (Sect. 4.3), a power-recycled Michelson with Fabry-Perot arm cavities (Sect. 4.4), and a Dual-recycled Michelson with Fabry-Perot arm cavities (Sect. 4.5). In each of the configurations, the effect of injected squeezed states is also explored. Section 4.6 concludes the chapter, discussing some of the assumptions used in calculations of quantum noise, as well as implications of injected squeezing towards strain sensitivity.

## 4.1 Quantum Noise in Measurement

This section provides an introduction of the influence of quantum noise in measurement, as a result of the amplitude and phase quadrature uncertainties.



S.S.Y. Chua, *Quantum Enhancement of a 4 km Laser Interferometer Gravitational-Wave Detector*, Springer Theses,
DOI 10.1007/978-3-319-17686-4_4

### 4.1.1 Amplitude Quadrature Uncertainty—Radiation Pressure Noise

Photons transferring their momentum onto an incident object gives rise to a force acting on the object that causes mechanical motion. Aptly called radiation pressure [4], the greater the incident optical power, the greater the number of photons present, the greater the radiation pressure. However, uncertainty in the amplitude quadrature $\Delta X_1$ translates into uncertainty in the photon number at a particular time/measurement point. Therefore, the amplitude quadrature uncertainty gives rise to *radiation pressure noise*. This noise source (after normalisation to the signal) scales proportionally to the optical power.

$$\sigma_{\mathrm{rad}} \propto \sqrt{P_{\mathrm{opt}}} \tag{4.1}$$

In most optical experiments, radiation pressure noise is often masked by other noise sources, such as thermal noise. However, second generation and future gravitational-wave detectors are expected to be sensitivity-limited (in part) by radiation pressure noise.

### 4.1.2 Phase Quadrature Uncertainty—Photon Shot Noise (Shot Noise)

*Photon shot noise* arises from the uncertainty in the arrival time of photons to the measurement point and is the manifestation of the uncertainty in the phase quadrature $\Delta X_2$ in measurement. This noise source (after normalisation to the signal) scales inversely proportional to optical power, as the phase quadrature uncertainty increases with decreasing number of photons present, or decreasing optical power.

$$\sigma_{\mathrm{shot}} \propto 1/\sqrt{P_{\mathrm{opt}}} \tag{4.2}$$

Photon shot noise is often encountered as the limit for many (quantum) optical experiments, and is a broadband sensitivity limitation for gravitational-wave detectors. For the purpose of this thesis, photon shot noise will be referred to simply as *shot noise*.

### 4.1.3 The Standard Quantum Limit (SQL)

Radiation pressure noise and shot noise are both proportionally dependent on optical power, but in opposite regimes. The combination of these two noises are their quadrature sum, and is the Quantum Noise curve in Fig. 4.1

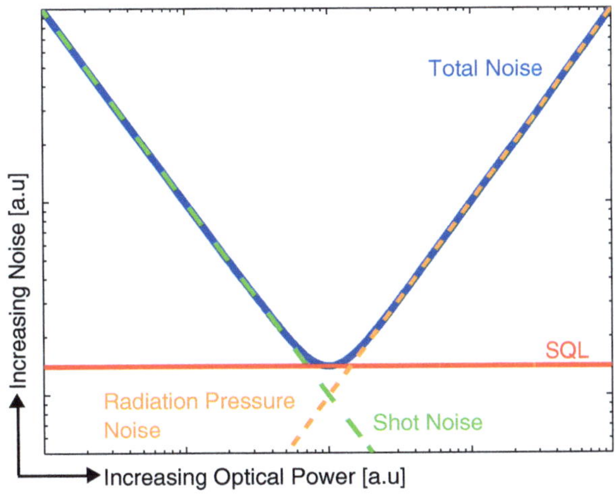


**Fig. 4.1** Quantum noise versus optical power. The sum of radiation pressure noise and shot noise contributions is shown by the Total Noise curve. The standard quantum limit is shown, intersecting with the Total Noise curve at an Optical Power level that has radiation pressure noise and shot noise contributions being equal

$$\sigma_{\text{total}} = \sqrt{\sigma_{\text{rad}}^2 + \sigma_{\text{shot}}^2} \propto \sqrt{\left(\frac{1}{P_{\text{opt}}}\right)^2 + P_{\text{opt}}^2} \tag{4.3}$$

In the low power regime, the shot noise contribution dominates the total quantum noise, while in the high power regime the radiation pressure noise contribution dominates. The optical power level where the contributions of radiation pressure noise and shot noise are equal coincides with the point where the quantum noise is at a minimum. It is this limit that is known as the *standard quantum limit* or *SQL*, and this is also shown in Fig. 4.1. Without applying quantum optical techniques such as squeezed states, the SQL is the minimum quantum noise that can be achieved. Therefore, the accuracy of amplitude and phase measurements is limited by the SQL.

## 4.2 Main Source of Quantum Noise in Gravitational-Wave Measurement

Gravitational-wave detectors are operated near a dark fringe, that is, the vast amount of the incident laser power is reflected by the Michelson back towards the laser [5]. As a consequence, the vast amount of classical laser noise and laser quantum noise is also reflected in this symmetric port. Complimentary to this, in the near-dark antisymmetric port of the Michelson, or *AS port*, the quantum noise is dominated by the incident vacuum state that is reflected by the interferometer back towards the readout photodetector. This is illustrated in Fig. 4.2a.

Caves [6] published the first complete argument of quantum noise in gravitational-wave detection readout arising from the vacuum field entering the AS port. This argument was reaffirmed by Braginsky et al. [7], who showed that the quantum limits in

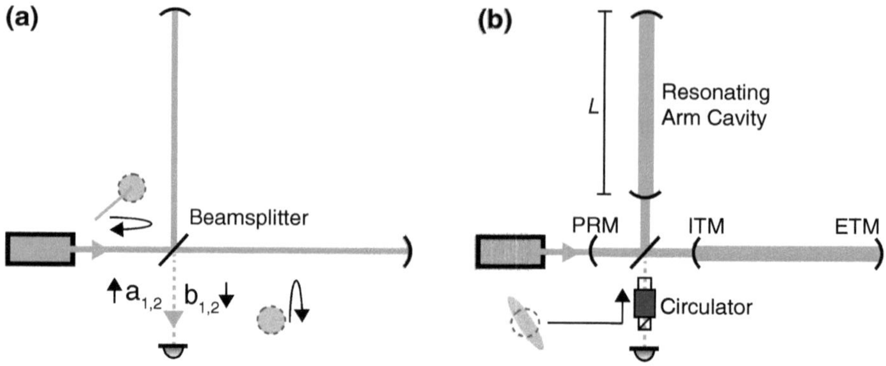


**Fig. 4.2** **a** A simple Michelson interferometer (Sect. 4.3). Quantum phasor diagrams depict the propagating direction of the input laser field and input vacuum field of the two ports of the Michelson. **b** Michelson with power-recycling mirror (*PRM*) and Fabry-Perot arm cavities (Sect. 4.4). Arm cavities are made up of an input test-mass (*ITM*) and end test-mass (*ETM*). The test masses are separated by length $L$. The output optical path shows squeezing injection via an optical circulator that spatially separates input and returning beams. Reprinted with permission from [8], copyright (2014) by IOP Publishing

gravitational-wave detectors are from *only* the quantum noise in the electromagnetic field, and not the (quantum) test mass quantisation. Therefore, the incident vacuum fluctuations in the AS port is the main source for quantum noise in measurement of gravitational-wave detectors.

The other sources of quantum noise are due to vacuum fluctuations that enter the interferometer due to optical losses from non-perfect optics. These sources will be considered in Sect. 4.4.5.

## 4.3 Simple Michelson Interferometers

A Michelson interferometer is the basic configuration for gravitational-wave detectors (introduced in Sect. 2.2.1). The simple Michelson configuration has two end mirrors isolated and suspended on pendula, thus acting as quasi-free test masses to sense the effect of a passing gravitational wave [5]. This section introduces the quantum noise for a simple Michelson, and the effect of injected squeezing.

### *4.3.1 Input-Output Relations for Quantum Noise*

For the simple Michelson (shown in Fig. 4.2a), the output amplitude quadrature, $b_1$, and phase quadrature, $b_2$ is given by [1]

$$b_1 = \Delta b_1, \qquad \Delta b_1 = a_1 e^{2i\beta} \tag{4.4}$$

$$b_2 = \Delta b_2 + \sqrt{2\mathscr{K}}\frac{h}{h_{SQL}}e^{i\beta}, \quad \Delta b_2 = (a_2 - \mathscr{K}a_1)e^{2i\beta} \tag{4.5}$$

$$b_\zeta = b_1 \sin(\zeta) + b_2 \cos(\zeta) \tag{4.6}$$

where $\Delta b_1$ and $\Delta b_2$ are output quadrature fluctuations due to quantum noise, $a_1$ and $a_2$ are the amplitude and phase quadratures of the field entering the interferometer's AS port, and $h$ is the gravitational wave strain. As illustrated, the field entering the AS port of an interferometer is the vacuum state. With a detection angle of $\zeta$, the corresponding output quadrature $b_\zeta$ can be measured. The single-pass phase shift of the field in the Michelson arm at sideband frequency $\Omega$ is

$$\beta = \arctan\left(2\Omega L/c\right) \tag{4.7}$$

where $L$ is the arm length and $c$ is the speed of light. The radiation-pressure back-action coupling constant $\mathscr{K}$ is given by [1]

$$\mathscr{K} = \frac{4I_0\omega_0}{mc^2\Omega^2} \tag{4.8}$$

where $I_0$ is the incident laser power on the interferometer beamsplitter, $\omega_0$ is the optical carrier frequency and $m$ the mass of the arm mirrors. Lastly,

$$h_{SQL} = \sqrt{\frac{4\hbar}{m\Omega^2 L^2}} \tag{4.9}$$

is the standard quantum limit for the square root of the gravitational-strain single-sided power spectral density [1]. This is the manifestation of the SQL described in Sect. 4.1.3 in a simple Michelson interferometer. As in Eq. 4.5, of primary importance is the output phase quadrature $b_2$, as it contains the gravitational wave signal. The quantum noise contribution ($\Delta b_2$) consists of a shot noise component, from input phase quadrature noise $a_2$, and a radiation pressure noise component, from input amplitude quadrature noise $a_1$ scaled by $-\mathscr{K}$. The single-sided power spectral density can be described by [1]

$$S_h = \frac{h_{SQL}^2}{2}\left(\frac{1}{\mathscr{K}} + \mathscr{K}\right) \tag{4.10}$$

and the strain sensitivity is $h = \sqrt{S_h}$. Figure 4.3a shows the strain sensitivity of a simple Michelson interferometer, normalised to the strain-equivalent SQL ($h_{SQL}(\Omega)$), with traces for radiation pressure and shot noise contributions, the SQL and the total quantum noise. Increasing or decreasing the input laser power will map the total quantum noise curve to the SQL at different sideband frequencies. It should be noted that Fig. 4.3a illustrates strain sensitivity versus sideband frequency, thus it is a spectrum.

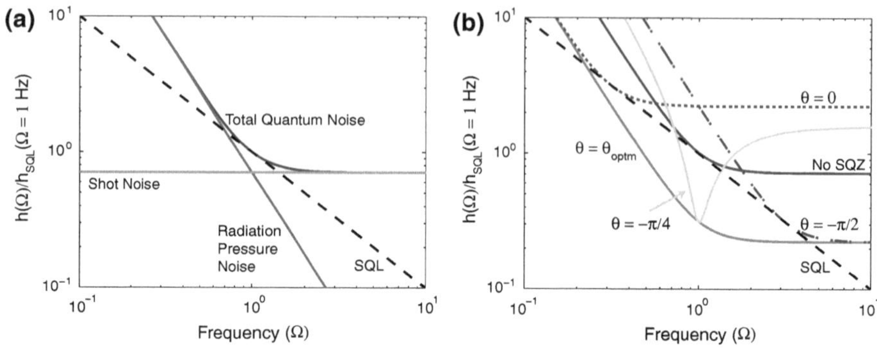


**Fig. 4.3** **a** Quantum noise limited strain sensitivity of a simple Michelson interferometer. **b** Simple Michelson interferometer with injected squeezing into the AS port, with different squeezing angles $\theta$. With squeezing, the SQL can be surpassed. Values used to calculate are $\omega_0 = 1.77 \times 10^{15}$ rad/s, $I_0 = 1 \times 10^4$ W, $m = 40$ kg, $L = 3995$ m. All traces are normalised to the SQL ($h_{SQL}(\Omega)$)

This is different to Fig. 4.1 that illustrates total quantum noise versus laser power, where the SQL represents the lower level limit of quantum noise.

### 4.3.2 *Squeezing in Simple Michelson Interferometers*

Quantum noise can be reduced by replacing the AS port input vacuum state with an injected squeezed state [9]. With an appropriate squeezing angle $\theta$, an injected squeezed state can reduce the quantum noise below the SQL [10, 11].

The power spectral density of a simple Michelson interferometer with injected squeezing is given by [1]

$$S_{\text{h-SQZ}} = \frac{h_{SQL}^2}{2}\left(\frac{1}{\mathcal{K}} + \mathcal{K}\right)\left[\cosh 2R - \cos\left(2(\theta + \Phi)\right)\sinh 2R\right] \quad (4.11)$$

where $R$ is the squeeze factor, and $\Phi = \text{arccot}(\mathcal{K})$. Figure 4.3b shows the strain sensitivity of a simple Michelson with injected squeezing at various squeezing angles. An injected phase squeezed state ($\theta = -\pi/2$) reduces the total quantum in the shot-noise limited regime, while increasing the noise in the radiation-pressure limited regime. This scenario is reversed for an amplitude squeezed state ($\theta = 0$). A squeezing angle of $\theta = -\pi/4$ allows for increased sensitivity below the SQL, but decreased sensitivity in both shot and radiation-pressure limited regimes.

To achieve optimal improvement from injected squeezing across all sideband frequencies, a *frequency-dependent squeezed state* is needed. This would entail an amplitude squeezed state ($\theta = 0$) in the radiation-pressure-noise-limited low frequency regime, rotated to $\theta = -\pi/4$ around the SQL frequency, then rotated further to a phase squeezed state $\theta = -\pi/2$ in the shot-noise-limited high frequency regime. The optimal squeezing angle is given by [1]

$$\theta_{\text{optm}}(\Omega) = -\text{arccot}(\mathcal{K}) \tag{4.12}$$

With optimal squeezing angle, Eq. 4.11 simplifies to

$$S_{\text{h-SQZoptm}} = \frac{h^2_{SQL}}{2}\left(\frac{1}{\mathcal{K}} + \mathcal{K}\right)e^{-2R} \tag{4.13}$$

which is Eq. 4.10 multiplied by $e^{-2R}$. This optimal strain sensitivity $\sqrt{S_{\text{h-SQZoptm}}}$ is shown in Fig. 4.3b. The injection of squeezing can be achieved with an optical circulator, shown in Fig. 4.2b. The workings of the optical circulator will be explored in Chap. 9.

## 4.4 Power-Recycled Fabry-Perot Michelson Gravitational-Wave Detectors

The Michelson interferometer with power-recycling and Fabry-Perot arm cavities is the configuration used in first generation and Generation 1.5 gravitational-wave detectors (Sect. 2.2.2). Known as Power-recycled Michelsons for the rest of this thesis, this configuration was used in the LIGO Squeezed Light Injection Experiment.

### *4.4.1 Additions to the Simple Michelson Configuration*

A Power-recycled Michelson is shown in Fig. 4.2b. The additions to the simple Michelson configuration are the power-recycling mirror (PRM) and optical arm cavities, formed between the input test-masses (ITM) and end test-masses (ETM).

The reflected power in the symmetric port from the Michelson (operating near a dark fringe) can be utilised by reflecting it back towards the Michelson. To 'recycle' the optical power, a power-recycling mirror (PRM) is placed in the symmetric port. This forms an optical cavity between itself and the reflecting Michelson, resulting in a resonant enhancement of the optical power incident on the interferometer beam-splitter.

The addition of optical cavities in the Michelson arms primarily allows an enhancement of the effect of a gravitational-wave signal [5]. The optical power is stored within each arm, resonantly building and effectively sampling the position of the ETM many times. This multiplies up the effect of the change in test mass position due to passing gravitational waves.

### 4.4.2 *Quantum Noise in a Lossless Power-Recycled Michelson*

The input-output relations of the quantum noise in this interferometer configuration are similar to the simple Michelson relations (Eqs. 4.4 and 4.5), except with $\beta$, $\mathcal{K}$ and $h_{SQL}$ being modified. These modified terms are denoted with superscript (I) to distinguish them from their simple Michelson counterparts. The single-pass phase shift of the field in the Michelson arm *cavity*, at sideband frequency $\Omega$, is now

$$\beta^I = \arctan(\Omega/\gamma) \tag{4.14}$$

where $\gamma = Tc/(4L)$ is the arm cavity half-width-half-maximum (half of the arm cavity's FWHM), and $T$ is the transmission of the input test-mass. The radiation-pressure back-action coupling constant becomes

$$\mathcal{K}^{\mathrm{I}} = \frac{2(I_0/I_{SQL})\gamma^4}{\Omega^2(\gamma^2+\Omega^2)} \tag{4.15}$$

where $I_{SQL}$ is the beamsplitter incident power needed to reach the SQL, given by

$$I_{SQL} = \frac{mL^2\gamma^4}{4\omega_0} \tag{4.16}$$

The strain at the standard quantum limit is now given by

$$h^{\mathrm{I}}_{SQL} = \sqrt{\frac{8\hbar}{m\Omega^2L^2}} \tag{4.17}$$

The strain sensitivity is thus given by

$$h^{\mathrm{I}} = \sqrt{S^{\mathrm{I}}_h} = \sqrt{\frac{(h^{\mathrm{I}}_{SQL})^2}{2}\left(\frac{1}{\mathcal{K}^{\mathrm{I}}} + \mathcal{K}^{\mathrm{I}}\right)} \tag{4.18}$$

which is square-root of the power spectral density, Eq. 4.10, with $\mathcal{K} \rightarrow \mathcal{K}^{\mathrm{I}}$, $h_{SQL} \rightarrow h^{\mathrm{I}}_{SQL}$. Figure 4.4a shows the total quantum noise, SQL and contributions from radiation pressure and shot noise for a Power-recycled Michelson.

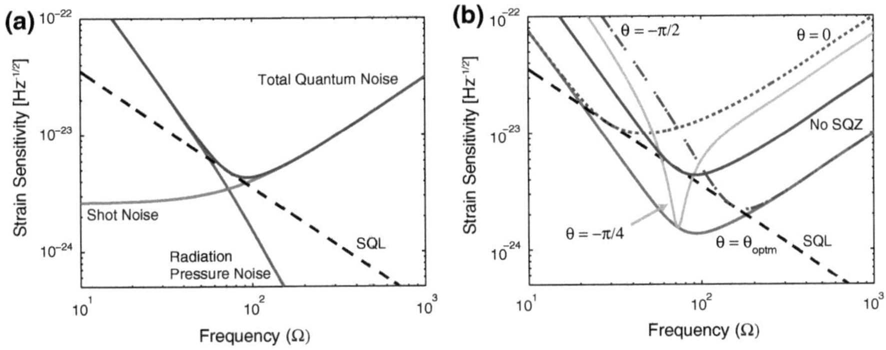


**Fig. 4.4** **a** Strain sensitivity of a Power-recycled Michelson, calculated using parameters found in Table 4.1 and **b** Strain sensitivity of a Power-recycled Michelson with injected squeezing at various squeezing angles

**Table 4.1** Parameters and values used in quantum noise calculations for the Power-recycled Michelson

| Parameter | Symbol | Value | Units |
|---|---|---|---|
| Laser Frequency | $\omega_0$ | $1.77 \times 10^{15}$ | rad/s |
| Detection Fourier frequency | $\Omega$ | 10–10,000 | Hz |
| Detection phase | $\zeta$ | $\pi/2$ | rad |
| Photodetection loss | $\lambda_{PD}$ | 0.1 | – |
| Laser power at beamsplitter | $I_0$ | 1200 | W |
| Mirror mass | $m$ | 10.7 | kg |
| ITM transmission | $T$ | 0.028 | – |
| Arm length | $L$ | 3995 | m |
| Arm cavity half-bandwidth | $\gamma$ | $Tc/(4L) \approx 2\pi \times 84$ Hz | rad |
| Arm cavity round trip loss (per cavity) | $\lambda_{AC}$ | $150 \times 10^{-6}$ | – |
| Arm cavity loss coefficient | $\varepsilon$ | $2\lambda_{AC}/T$ | – |
| Fractional photon loss in arm cavities | $\mathcal{E}$ | $2\varepsilon/(1 + (\Omega/\gamma)^2)$ | – |
| Fractional photon loss in output train | $\mathcal{E}_{OT}$ | 0.005 | – |

Apart from the fractional photon loss in output train $\mathcal{E}_{OT}$, these are the Enhanced LIGO interferometer parameters

### 4.4.3 Arm Cavities and Power-Recycling Effect on Quantum Noise

The presence of arm cavities effects the quantum noise limited sensitivity in two ways. Shown in Fig. 4.4a the first effect is the shot noise limited sensitivity is now shaped. This is from the arm cavity half-linewidth $\gamma$, entering the equation via the $\mathscr{K}^I$ term. This shaping will be expanded on in Sect. 4.6.1. The second effect is that the standard quantum limit is now $\sqrt{2}$ times larger than the simple Michelson case. This arises from each arm cavity (with two test masses) acting as an equivalent single test mass with reduced mass $m \rightarrow m/2$ (see footnote 3 of Ref. [1]).

Power-recycling resonantly enhances the optical power incident on the interferometer beamsplitter. This resonant enhancement acts on the quantum noise limit of the interferometer via increasing the power incident on the interferometer beamsplitter, $I_0$.

### 4.4.4 Squeezing Enhancement of a Lossless Power-Recycled Michelson

Figure 4.4b shows the enhancements possible with the injection of squeezing. These traces are calculated identically to the simple Michelson case (Sect. 4.3.2), except with the optimal squeezing angle now calculated with the modified radiation-pressure back-action coupling constant (Eq. 4.12 with $\mathscr{K}^{\mathrm{I}}$). Depending on the squeezing ellipse angle, the improvement will effect different quantum noise regimes.

### 4.4.5 Power-Recycled Michelson with Optical Losses

So far, the analysis of quantum noise has proceeded under lossless conditions. The presence of loss couples vacuum fluctuations into a system. With losses, the input-output relations are further modified, with terms denoted with subscript $(*)$. The input-output relations are now [1]

$$b_1 = \Delta b_1, \quad b_2 = \Delta b_2 + \sqrt{2\mathscr{K}_{(*)}} \frac{h}{h^{\mathrm{I}}_{SQL}} e^{i\beta_*} \tag{4.19}$$

The single-pass phase shift of the field in the Michelson arm cavity is now

$$\beta_{(*)} = \tan^{-1}\left(\frac{\Omega/\gamma}{1+\varepsilon/2}\right) \tag{4.20}$$

where $\varepsilon$ is the arm cavity loss coefficient. The radiation-pressure back-action coupling constant is modified to

$$\mathscr{K}_{(*)} = \mathscr{K}^{\mathrm{I}}\left(1 - \frac{1}{2}\mathscr{E}\right) \tag{4.21}$$

with the fractional power loss in the arm cavities $\mathscr{E}$ given by

$$\mathscr{E} = \frac{2\varepsilon}{1 + (\Omega/\gamma)^2} \tag{4.22}$$

The power spectral density of the Power-recycled Michelson with optical losses can be written as [1]

$$S_{\mathrm{hL}}^{\mathrm{I}} = \frac{(h_{SQL}^{\mathrm{I}})^2}{2}\left[\underbrace{\frac{\mathscr{E} + \mathscr{E}_{OT}}{\mathscr{K}^{\mathrm{I}}}}_{\mathrm{i}} + \underbrace{\frac{\varepsilon}{2}\mathscr{K}^{\mathrm{I}}}_{\mathrm{ii}} + \underbrace{\left(1 - \frac{1}{2}\mathscr{E}\right)\left(\frac{1}{\mathscr{K}^{\mathrm{I}}} + \mathscr{K}^{\mathrm{I}}\right)}_{\mathrm{iii}}\right] \tag{4.23}$$

where $\mathscr{E}_{OT}$ is the sum of the fractional photon losses in the interferometer output in the measurement. $\mathscr{E}_{OT}$ includes fractional loss from the circulator and readout photodetector.

The three components of $S_{\mathrm{hL}}^{\mathrm{I}}$ are:

i. The coupling of vacuum fluctuations due to optical loss in the arm cavities and the interferometer optical output train. The entrance of these vacuum fluctuations affect the shot noise contribution to quantum noise.
ii. The backaction coupling of vacuum fluctuations due to optical loss in the arm cavities. This results in amplitude quadrature noise contributing to the measurement. The entrance of these vacuum fluctuations affect the radiation pressure contribution to quantum noise.
iii. Vacuum fluctuations entering the AS port of the interferometer, decreased by the optical losses in the reflecting Michelson.

As expected, if the loss terms in Eq. 4.23 were zero, we return to the lossless power spectral density (Eq. 4.18). Thus with small fractional losses ($\varepsilon \ll 1$), calculations of the strain sensitivity and power spectral density can be well approximated by the Power-recycled Michelson lossless equations.

### *4.4.6 Squeezing in a Power-Recycled Michelson with Optical Losses*

For completeness, the strain power spectral density for a Power-recycled Michelson with losses is stated below [1]

$$S^{\mathrm{I}}_{\mathrm{hL-SQZ}} = \frac{(h^{\mathrm{I}}_{SQL})^2}{2}\left[\frac{\mathscr{E}+\mathscr{E}_{OT}}{\mathscr{K}^{\mathrm{I}}} + \frac{\varepsilon}{2}\mathscr{K}^{\mathrm{I}} + \left(1-\frac{1}{2}\mathscr{E}\right)\left(\frac{1}{\mathscr{K}^{\mathrm{I}}}+\mathscr{K}^{\mathrm{I}}\right)\right.$$
$$\left.\times\left\{\cosh 2R - \cos\left(2(\theta+\Phi)\right)\sinh 2R\right\}\right] \tag{4.24}$$

With optimal squeezing angle, the power spectral density becomes [1]

$$S^{\mathrm{I}}_{\mathrm{hL-SQZopt}} = \frac{(h^{\mathrm{I}}_{SQL})^2}{2}\left[\frac{\mathscr{E}+\mathscr{E}_{OT}}{\mathscr{K}^{\mathrm{I}}} + \frac{\varepsilon}{2}\mathscr{K}^{\mathrm{I}} + \left(1-\frac{1}{2}\mathscr{E}\right)\left(\frac{1}{\mathscr{K}^{\mathrm{I}}}+\mathscr{K}^{\mathrm{I}}\right)e^{-2R}\right] \tag{4.25}$$

The effect of squeezing is very similar to the lossless case, but highlights the important distinction about the different vacuum noise components entering the system. The injection of squeezing *only aids in changing the quantum noise due to the vacuum fluctuations entering the AS port, and not other vacuum fluctuation sources.*

## 4.5 Dual-Recycled Fabry-Perot Michelson Gravitational-Wave Detectors

The 'Dual-recycled Michelson' interferometer extends the Power-recycled Michelson configuration with the addition of a mirror in the AS port. This mirror, called the signal recycling mirror (SRM), now forms an optical cavity between itself and the Michelson, resulting in a resonant enhancement of gravitational-wave signals. This is the technique of signal recycling [12]. The Dual-recycled Michelson configuration will be used in second generation gravitational-wave detectors (Sect. 2.2.3).

The Dual-recycled Michelson configuration is beyond the experiments contained within this thesis, however, the strain sensitivity curves are presented here for completeness. The quantum noise of a Dual-recycled Michelson with optical losses is shown in Fig. 4.5a, corresponding to quantum noise curve as plotted by the GWINC software of Fig. 4.6. The effect of injected squeezing at various squeezing angles is shown in Fig. 4.5b. The full quantum-noise equations for Dual-recycled Michelsons can be found in Appendix A.

## 4.6 Implications and Assumptions

This section discusses some of the assumptions used in the above calculations. The implications and limitations of injecting squeezed states are also discussed.

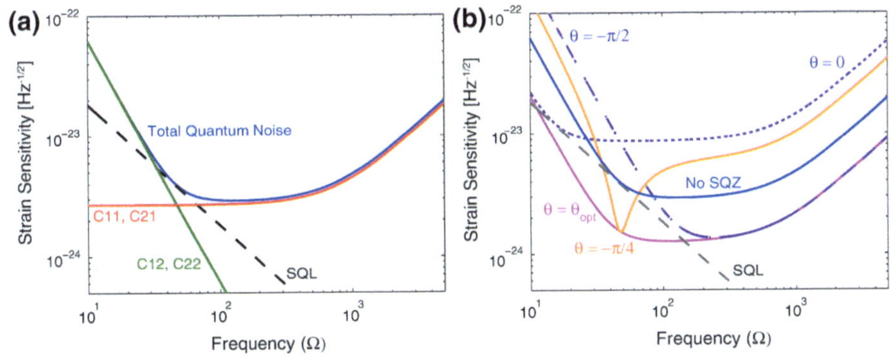


**Fig. 4.5** **a** Strain sensitivity of a Dual-recycled Michelson with optical losses. The contribution of the AS port vacuum fluctuations are denoted by the C—coefficient curves. Full equations are contained in Appendix A. The total quantum noise is the form plotted for the quantum noise contribution in GWINC v3 [13] (see Fig. 4.6 for comparison). **b** Strain sensitivity with injected squeezing at various squeezing angles

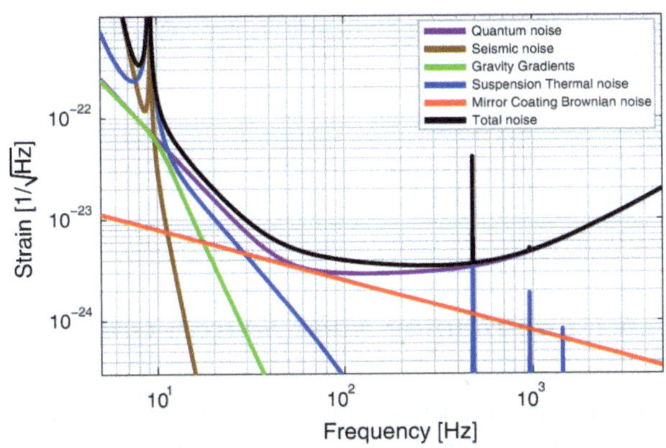


**Fig. 4.6** Figure 2.7 repeated—the predicted noise budget of Advanced LIGO, showing the contributions of the various limiting noise sources. Plot made using GWINC v3 [13]

## *4.6.1 Signal-to-Noise Ratio*

The shot noise is spectrally flat, as introduced in the shot noise section of Chap. 3. In a simple Michelson, the shot noise limited contribution is also spectrally flat. One might ask why the shot noise contribution is shaped for the Power-recycled and Dual-recycled Michelson configurations. This is resolved by a signal-to-noise argument.

The shaping is due to the presence of the arm cavities in the Michelson. The gravitational-wave signal is shaped by the arm cavity resonance profile, via $\mathscr{K}^1$ (see Eqs. 4.5 and 4.15). This attenuates the gravitational-wave signal at detection Fourier frequencies $\Omega$ away from resonance. With shot noise contribution at a constant level, but with reducing gravitational-wave signal as a function of frequency, the

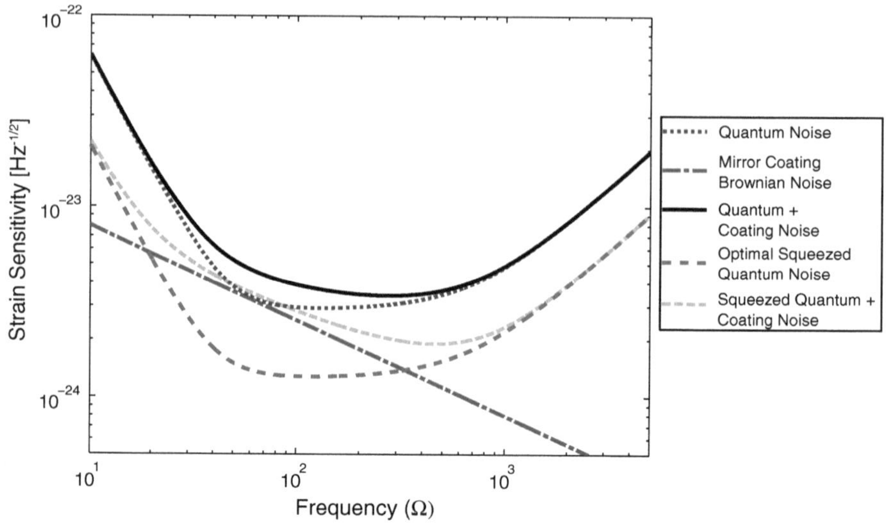


**Fig. 4.7** Strain sensitivity of a dual-recycled Michelson with applied squeezing, accounting for quantum noise and coating brownian thermal noise. The total improvement potential from injected squeezing is masked by the coating brownian thermal noise

gravitational-wave signal-to-noise ratio (hence the strain sensitivity) is reduced as a function of frequency.

### *4.6.2 Contributing Noise Sources Other Than Quantum Noise*

The chapter has focussed only on the impact of quantum noise in the interferometer readout. Figure 4.6 shows that quantum noise broadly limits the predicted strain sensitivity of second-generation detector Advanced LIGO. However, other noise from hardware and environment sources (Sect. 2.3) contribute in limiting the sensitivity.

Squeezing injection only affects the quantum noise contribution and does not reduce other noise contributions should they begin to dominate the quantum noise. For example, thermal Brownian noise of mirror coatings is a noise source comparable in contribution as the quantum noise around 100 Hz.[1] Figure 4.7 shows the result of coating thermal noise masking the sensitivity-improvement possible with optimal squeezing injection. Therefore, the full benefit of injected squeezing is only gained in regions where quantum noise is dominant.

[1]Coating thermal noise is an active area of research [14].

### 4.6.3 Squeezing Injection Optical Losses

The calculations for the quantum noise have also assumed perfect squeezing injection. However, optical losses degrade squeezing magnitude, returning the state back to the vacuum state. To factor in squeezing injection with optical loss, the effective squeezing magnitude should be reduced in a similar fashion. Optical loss in squeezing measurement will be discussed in Sect. 6.4.1.

### 4.6.4 Frequency Dependent Squeezing and the Heisenberg Uncertainty Principle

With the optimised squeezing angle cases, one might enquire whether if the Heisenberg Uncertainty Principle is being violated as there appears to be simultaneous improvement in shot noise and radiation-pressure noise limited regimes. The Heisenberg Uncertainty Principle is not violated by frequency dependent squeezed states, as such states satisfy the Uncertainty Principle *at each sideband frequency*. Should one measure the complimentary quadrature, the total noise curve would be increased (the strain sensitivity reduced) across all sideband frequencies.

### 4.6.5 Squeezing and Optical Filter Cavities

Frequency-dependent squeezing is needed in order to access the full benefit of squeezed-state injection. Observation of frequency-dependent squeezing has been experimentally realised at MHz frequencies [15], and suggestions for methods to obtain such squeezed states in the gravitational-wave detection band have been proposed.

Kimble et al. [1] first proposed to use optical cavities, named *filter cavities*, in the squeezing injection path in order to frequency-dependently rotate the squeezing ellipse. Theoretically, filter cavities provide an excellent method for squeezing rotation. Physically, however, filter cavities need to be lossless, so as to not degrade the incident squeezing, as well as be able to rotate the squeezing around the appropriate detection frequency where the radiation-pressure-noise limited and shot-noise limited regimes crossover (i.e. around the SQL intersection point). These requirements place stringent tolerances on the optical cavity, and are currently-active areas of investigation [16–18].

Another proposal is to use cavities as Amplitude filter cavities [19, 20]. This method does not give optimised frequency-dependent squeezing. Instead, the optical cavities are used to filter away the anti-squeezing component of the state. For phase squeezing that improves the shot noise regime, the amplitude anti-squeezing is filtered away. The radiation-pressure noise regime thus remains unaltered, at the vacuum fluctuation level. The advantage of using Amplitude filter cavities is that the requirements for these filter cavities are relaxed considerably, both in terms of cavity loss tolerances and cavity length.

## 4.7 Summary

This chapter introduces the calculations for obtaining the strain-sensitivity limited quantum noise traces for interferometers and gravitational-wave detectors. The cases of a simple Michelson, power-recycled Michelson with arm cavities, and the dual-recycled Michelson with arm cavities were explored. The effect of squeezing was covered for each of the cases. Lastly, the assumptions for the calculations and implications for squeezing injection into future gravitational-wave detectors were discussed.

## References

1. H.J. Kimble, Y. Levin, A.B. Matsko, K.S. Thorne, S.P. Vyatchanin, Conversion of conventional gravitational-wave interferometers into quantum nondemolition interferometers by modifying their input and/or output optics. Phys. Rev. D **65**, 022002 (2002)
2. A. Buonanno, Y. Chen, Quantum noise in second generation, signal-recycled laser interferometric gravitational-wave detectors. Phys. Rev. D **64**, 042006 (2001)
3. A. Buonanno, Y. Chen, Improving the sensitivity to gravitational-wave sources by modifying the input-output optics of advanced interferometers. Phys. Rev. D **69**, 102004 (2004)
4. B.E.A. Saleh, M.C. Teich, *Fundamentals of Photonics* (Wiley, New York, 1991)
5. P.R. Saulson, *Fundamentals of Interferometric Gravitational Wave Detectors* (World Scientific, Singapore, 1994)
6. C.M. Caves, Quantum-mechanical radiation-pressure fluctuations in an interferometer. Phys. Rev. Lett. **45**, 75–79 (1980)
7. V.B. Braginsky, M.L. Gorodetsky, F.Y. Khalili, A.B. Matsko, K.S. Thorne, S.P. Vyatchanin, Noise in gravitational-wave detectors and other classical-force measurements is not influenced by test-mass quantization. Phys. Rev. D **67**, 082001 (2003)
8. S.S.Y. Chua, B.J.J. Slagmolen, D.A. Shaddock, D.E. McClelland, Quantum squeezed light in gravitational-wave detectors. Class. Quantum Gravity **31**, 183001 (2014)
9. C.M. Caves, Quantum-mechanical noise in an interferometer. Phys. Rev. D, **23**(8) (1981)
10. W.G. Unruh, in *Quantum Optics, Experimental Gravitation and Measurement Theory*, ed. by P. Meystre, M.O. Scully (Plenum Press, New York, 1983), p. 647
11. M.T. Jaekel, S. Reynaud, Quantum limits in interferometric measurements. EPL (Europhys. Lett.) **13**(4) 301 (1990)
12. B.J. Meers, Recycling in laser-interferometric gravitational-wave detectors. Phys. Rev. D **38**, 2317 (1988)
13. Gravitational Wave Interferometer Noise Calculator (GWINC) v3, https://awiki.ligo-wa.caltech.edu/aLIGO/GWINC
14. G.M. Harry, A.M. Gretarsson, P.R. Saulson, S.E. Kittelberger, S.D. Penn, W.J. Startin, S. Rowan, M.M. Fejer, D.R.M. Crooks, G. Cagnoli, J. Hough, N. Nakagawa, Thermal noise in interferometric gravitational wave detectors due to dielectric optical coatings. Class. Quantum Gravity **19**, 897–999 (2002)

15. S. Chelkowski, H. Vahlbruch, B. Hage, A. Franzen, N. Lastzka, K. Danzmann, R. Schnabel, Experimental characterization of frequency-dependent squeezed light. Phys. Rev. A **71**, 013806 (2005)
16. J. Harms, Y. Chen, S. Chelkowski, A. Franzen, H. Vahlbruch, K. Danzmann, R. Schnabel, Squeezed-input, optical-spring, signal-recycled gravitational-wave detectors. Phys. Rev. D **68**, 042001 (2003)
17. F.Y. Khalili. Optimal configurations of filter cavity in future gravitational-wave detectors. Phys. Rev. D **81** (2010)
18. F. Magana-Sandoval, R. Adhikari, V. Frolov, J. Harms, J. Lee, S. Sankar, P.R. Saulson, J.R. Smith, Large-angle scattered light measurements for quantum-noise filter cavity design studies. J. Opt. Soc. Am. A **29**(8), (2012)
19. T. Corbitt, N. Mavalvala, S.E. Whitcomb, Optical cavities as amplitude filters for squeezed fields. Phys. Rev. D **70**, 022002 (2004)
20. F.Y. Khalili, Increasing future gravitational-wave detectors' sensitivity by means of amplitude filter cavities and quantum entanglement. Phys. Rev. D **77**, 062003 (2008)

# Chapter 5
# Squeezed State Generation for Gravitational-Wave Detection

**Abstract** The Heisenberg Uncertainty Principle manifests itself in many different parameter-pairs. Squeezed states of those parameter-pairs and their diverse generating apparatus have been and are being developed. Examples include number squeezing with atoms (Gross et al., Nature 464:1165–1169, 2010, [1]), atomic spin squeezing (Hamley et al., Nature Phys. 8:305–308, 2012, [2]), optical polarisation squeezing (Korolkova et al., Phys Rev A 65:052306, 2002; Corney et al., Simulations and experiments on polarisation squeezing in optical fibre, 2008, [3, 4]), and ponderomotive squeezing (Corbitt et al., Phys Rev A 73:023801, 2006; Marino et al., Phys Rev Lett, 2010; Brooks et al., Nature 488:476–480, 2012, [5–7]). This chapter reviews the type of squeezed vacuum generator that is currently applicable for gravitational-wave detector enhancement. A simplified overview of a squeezed state generator is covered, with key components and physicals processes treated in detail. The recommended sources for further detail are the works of Buchler (Electro-optic control of quantum measurements, 2008, [8]), McKenzie (Squeezing in the audio gravitational wave detection band, 2008, [9]) and Vahlbruch (Squeezed light for gravitational wave astronomy, 2004, [10]).

## 5.1 Squeezed Light Source Overview

The type of squeezed state generator, or *squeezer*, currently applicable to gravitational-wave detectors uses nonlinear optical materials to manipulate the vacuum quadrature noise properties. These 'crystal squeezers' are also used for other quantum optical experiments, such as entanglement [11] and teleportation [12].

Figure 5.1 shows a simplified schematic of a squeezer. The main aspects of a squeezer, described in detail in the subsequent sections, are:

1. Laser input stages: The two source lasers (Red sections)
2. Nonlinear optical stages: The Second Harmonic Generator and Optical Parametric Oscillator (Green sections)
3. Detection stage: The Balanced Homodyne Detector (Purple section)



S.S.Y. Chua, *Quantum Enhancement of a 4 km Laser Interferometer Gravitational-Wave Detector*, Springer Theses,
DOI 10.1007/978-3-319-17686-4_5

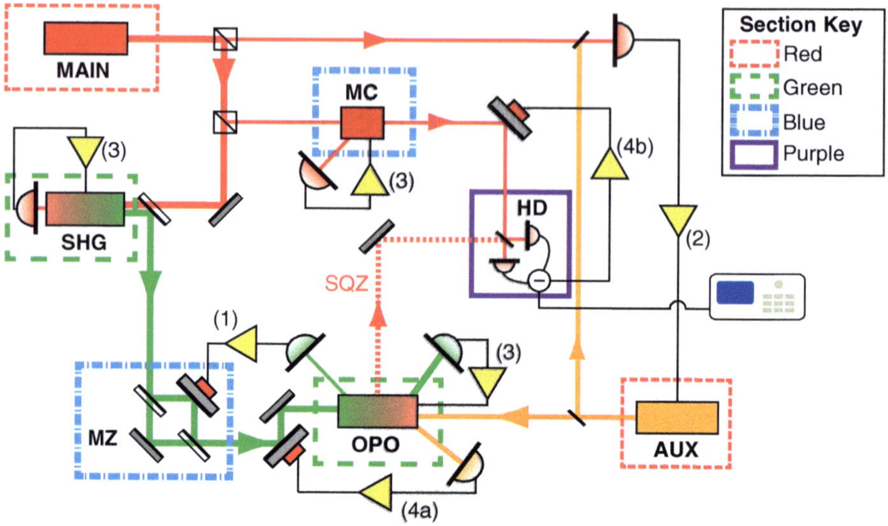


**Fig. 5.1** A simplified schematic of a squeezed light source. The *dashed* optical beam 'SQZ' represents the generated squeezed vacuum. *MAIN* and *AUX* the "Main" and "Auxiliary" source lasers; *SHG* Second harmonic generator; *OPO* Optical parametric oscillator; *HD* Homodyne detector; *MC* Mode cleaner; *MZ* Mach-Zehnder interferometer. The control loops are *(1)* DC voltage subtraction locking, *(2)* Offset phase locking, *(3)* Pound-Drever-Hall locking and *(4)* Coherent sideband locking

4. Filtering and stabilisation stages: The Mode Cleaner and Mach-Zehnder Interferometer (Blue sections)
5. Control loop stages: DC voltage subtraction locking, offset frequency locking, Pound-Drever-Hall locking and Coherent Sideband locking (Feedback arrows)

As a walkthrough, we begin at the main laser (MAIN—Sect. 5.2), the main source of laser light for the experiment. Most of the main laser light is sent to the Second Harmonic Generator (SHG—Sect. 5.3.3), while a small tapoff beam, after spatial mode filtering by the mode cleaner (MC—Sect. 5.5.1) is used as part of the Balanced Homodyne Detector (HD—Sect. 5.4.2). At the Second Harmonic Generator, the main laser light is frequency doubled using a second order nonlinear interaction process to produce second harmonic light. This second harmonic light is then sent, via a Mach-Zehnder interferometer to stabilise the second harmonic optical power (MZ—Sect. 5.5.2), to the Optical Parametric Oscillator (OPO—Sect. 5.3.4). At the Optical Parametric Oscillator, another second order nonlinear interaction process, driven by the second harmonic light, generates the squeezed vacuum state. The squeezed vacuum is then sent to the Homodyne Detector to be measured.

The DC voltage subtraction technique [Loop (1)—Sect. 5.6.1] is used to control the Mach-Zehnder arm length. The offset frequency locking loop [Loop (2)—Sect. 5.6.2] controls the relative frequency offset between the main and auxiliary (AUX) lasers. The Pound-Drever-Hall locking technique [Loops (3)—Sect. 5.6.3] is used to maintain optical cavities on resonance. The auxiliary laser is sent to the

Optical Parametric Oscillator so that it forms the Coherent Sideband locking field [Loops (4a) and (4b)] to control the relative phase of the squeezed state to the Homodyne Detector Local Oscillator phase. The Coherent Sideband locking technique will be explored in Chap. 6.

## 5.2 Laser Input Stages

The fields from the two lasers are the main inputs that are used to drive the entire squeezed light apparatus. The light from the main laser (MAIN) is predominantly used as the input for the Second Harmonic Generator, with the remainder providing the Local Oscillator field for the Homodyne Detector, and a low-power field for optimising the Optical Parametric Oscillator. The optical frequency of the main laser field sets the optical frequency of the produced squeezed field, and is named the *fundamental field* of the squeezer. The auxiliary laser (AUX) is a second laser that is frequency shifted from the fundamental field, and is the field used in the Coherent Sideband locking technique for the readout and control of the output phase of the squeezed state.

## 5.3 Nonlinear Optical Stages

The nonlinear optical stages are the key components that define a squeezed-light generator. This section introduces nonlinear optical interactions, focussing on the second order ($\chi_2$) nonlinear process. The Second Harmonic Generator and Optical Parametric Oscillator stages of a squeezer are described in detail.

### *5.3.1 Nonlinear Optical Interactions*

Nonlinear optical interactions are identifiable by the response of a material (system) to an applied optical field that depends in a nonlinear manner upon the amplitude of the applied field [13]. In dielectric media, the response of the material to an incident optical field $E$ is an induced polarisation $\mathscr{P}$ that can be written as

$$\mathscr{P} = \varepsilon_0(\chi_1 E + \chi_2 E^2 + \chi_3 E^3 + \cdots) \tag{5.1}$$

where the $\chi_n$ represent the $n$th order nonlinear coefficient, and $\varepsilon_0$ is the permittivity of free space. We focus on the second order nonlinear process, namely the dominant coefficient is the $\chi_2$ term.

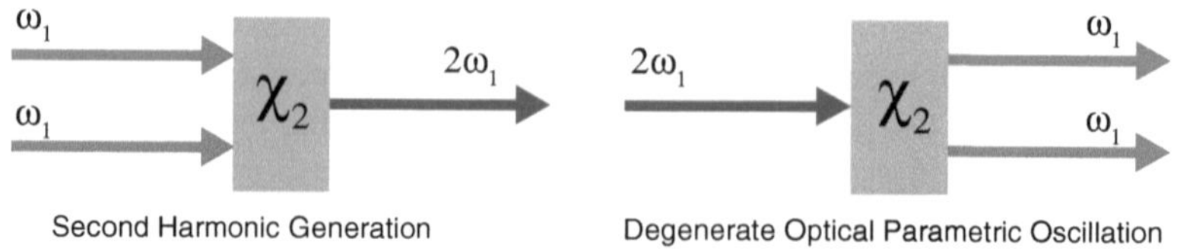


**Fig. 5.2** Illustration of the two degenerate $\chi_2$ nonlinear optical processes

#### 5.3.1.1 Second Order Nonlinear Optical Interactions

Second-order (or $\chi_2$) nonlinear optical interactions form part of the family of *three-wave mixing* processes [14]. Two such $\chi_2$ processes are:

- Sum frequency generation: two low energy photons ($\omega_1, \omega_2$) are converted into a high energy photon ($\omega_3$) that satisfy the condition $\omega_1 + \omega_2 = \omega_3$. When $\omega_1 = \omega_2$, the process is called *second harmonic generation*
- Optical parametric oscillation: A high energy photon ($\omega_1$) is converted into two low energy photons ($\omega_2, \omega_3$) that satisfy the condition $\omega_1 = \omega_2 + \omega_3$. When $\omega_2 = \omega_3$, the process is called *degenerate optical parametric oscillation*

We only explore the degenerate cases of each of the above processes (second harmonic generation and degenerate optical parametric oscillation), as they are the key processes for squeezed light generation in the squeezer. These specific nonlinear processes are illustrated by Fig. 5.2.

#### 5.3.1.2 Phase Matching

The relationship between the high and low energy photons partaking in the interaction process must satisfy the Law of Conservation of Energy, that is

$$\begin{aligned}\sum_{in} E_i = \sum_{out} E_j \Rightarrow \sum_{in} \hbar\omega_i &= \sum_{out} \hbar\omega_j \\ \Rightarrow \sum_{in} \omega_i &= \sum_{out} \omega_j\end{aligned} \tag{5.2}$$

Each field of frequency $\omega_{i,j}$ also has a corresponding wave vector $\mathbf{k_{i,j}}$, thus the Law of Conservation of Momentum also must be satisfied.

$$\hbar \sum_{in} \mathbf{k_i} = \hbar \sum_{out} \mathbf{k_j} \tag{5.3}$$

Assuming that input and output fields of the interaction process are propagating collinearly with each other, the two conservation principles imply that the input and output fields must propagate at the same speed through the nonlinear media. This

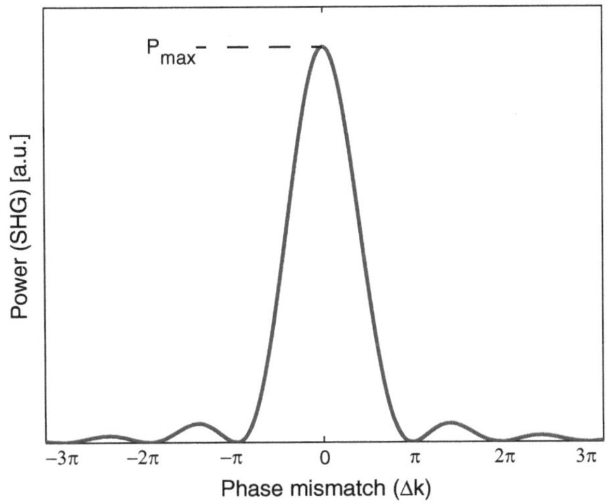


**Fig. 5.3** Change in generated second-harmonic power as a function of phase mismatch $\Delta k$. The nulls of the Sinc-squared function (for a fixed length) appear when the phase mismatch $\Delta k = n\pi$ where $n$ is an integer

gives rise to the 'phase matching' parameter that needs to be satisfied for efficient nonlinear conversion to occur. To illustrate, we define the phase mismatch term $\Delta k$ as the difference between input and generated field wave-vector magnitudes

$$\Delta k = \sum_{in} k_i - \sum_{out} k_j \tag{5.4}$$

The expression for the generated field power versus phase mismatch for the sum-frequency generation case is given by [13]

$$P_{\text{sum-freq}} = P^{max}_{\text{sum-freq}} \times \text{sinc}^2(\Delta k z/2) \tag{5.5}$$

Figure 5.3 shows Eq. 5.5 as a function of phase mismatch. Moving away from perfect phase matching ($\Delta k = 0$) results in a large decrease in the efficiency of the nonlinear process. Perfect phase-matching is often difficult to naturally achieve in materials. This is due to, for example, material dispersion—a material's refractive index varying as a function of optical frequency. However, there are two main techniques used to achieve good phase matching:

- **Birefringent Phase Matching**—Birefringence is the phenomenon where the refractive index of a material is dependent on the polarisation direction of the incoming field. There are two categories of Birefringent Phase Matching, Type I and Type II. These types are differentiated by the low energy photons being both polarised to the material ordinary axis (denoted by "→"), or one in each of the extraordinary (denoted by "↑") and ordinary axes, that is

$$\text{Type I}: \quad \uparrow_{\omega_3} = \rightarrow_{\omega_1} + \rightarrow_{\omega_2} \tag{5.6}$$
$$\text{Type II}: \quad \uparrow_{\omega_3} = \uparrow_{\omega_1} + \rightarrow_{\omega_2} \tag{5.7}$$

Example materials include Type I magnesium-oxide-doped lithium niobate (MgO: $LiNbO_3$) and Type II potassium titanyl phosphate ($KTiOPO_4$, known commonly as KTP).

- **Quasi-phase matching**—Quasi-phase matching [15] is a technique where the nonlinear coefficient is inverted periodically by modulating the material crystal domain structures. This is referred to as periodic poling and this technique counters the effect of material dispersion. Therefore it allows for the phase of the newly generated field to be quasi-phase matched with existing generated field. Although not as conversion-process efficient as the non-poled materials, quasi-phase-matched materials have an advantage that input/output fields are in the same polarisation, that is

$$\uparrow_{\omega_3} = \uparrow_{\omega_1} + \uparrow_{\omega_2} \tag{5.8}$$

This allows access to larger nonlinear coefficients inaccessible to birefringent phase-matched materials. Examples of quasi-phase-matched materials include periodically-poled KTP (PPKTP) and periodically-poled lithium niobate (PPLN) [16].

### 5.3.2 $\chi_2$ Nonlinear Cavity Quantum Langevin Equations

The $\chi_2$ nonlinear process can be enhanced by placing the nonlinear media into an optical cavity. To describe this system, the cavity Quantum Langevin equations from Sect. 3.6 are now modified. Figure 5.4 shows a generic cavity containing within it a nonlinear medium. We define similar quantities for the inputs and outputs fields, with the specific addition of the harmonic $(B, b)$ fields, while maintaining the $(A, a)$ notation for the fundamental frequency fields.

For the remainder of this thesis, the harmonic field will be referred to as the *pump field*. We also make a distinction that if there is a bright coherent field at the fundamental frequency incident on the OPO, it will be referred to as the *seed field*.

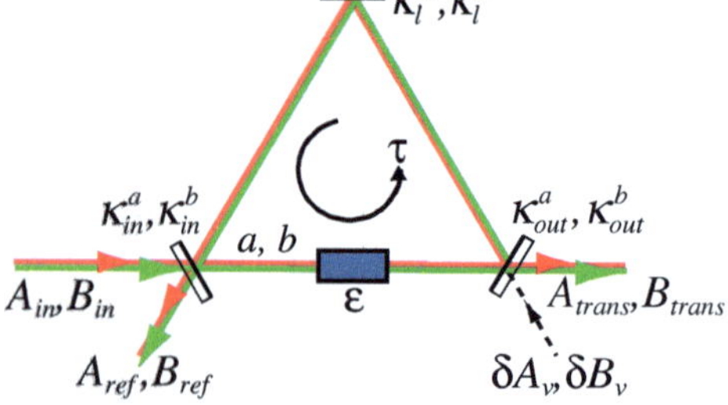


**Fig. 5.4** Optical cavity with nonlinear medium. This allows for enhancement of nonlinear processes via the interaction of resonantly-enhanced optical fields. $\varepsilon$ is the nonlinear coupling coefficient

The nonlinear quantum Langevin equations of motion for the fundamental and pump fields are defined as [13]

$$\dot{a} = -(\kappa_a + i\Delta_a)a + \varepsilon a^\dagger b + \sqrt{2\kappa_{in}^a}A_{in} + \sqrt{2\kappa_{out}^a}\delta A_v + \sqrt{2\kappa_l^a}\delta A_l \quad (5.9)$$

$$\dot{b} = -(\kappa_b + i\Delta_b)b - \frac{\varepsilon}{2}a^2 + \sqrt{2\kappa_{in}^b}B_{in} + \sqrt{2\kappa_{out}^b}\delta B_v + \sqrt{2\kappa_l^b}\delta B_l \quad (5.10)$$

The key characteristic is the interdependence between the fundamental and pump fields. The terms containing $\varepsilon$, the nonlinear coupling coefficient, encapsulate the interdependence, accounting for the flow of photons between one field to the other. As in Sect. 3.6, these equations can be similarly simplified by assuming that the cavity is held on resonance ($\Delta_a = \Delta_b = 0$).

### 5.3.3 Second Harmonic Generator (SHG)

The *Second Harmonic Generator* or *SHG* is a device that resonantly converts photons from the fundamental frequency into photons at the pump frequency, via the degenerate sum frequency generation process.

In the classical regime, the Langevin equations simplify as fluctuation terms are not considered and quantum operators become their expectation values ($a \rightarrow \langle a\rangle,\ b \rightarrow \langle b\rangle$). The classical cavity equations are

$$\begin{aligned}\langle\dot{a}\rangle &= -\kappa_a\langle a\rangle + \varepsilon\langle a\rangle\langle b\rangle + \sqrt{2\kappa_{\text{in}}^a}A_{\text{in}} \\ \langle\dot{b}\rangle &= -\kappa_b\langle b\rangle - \frac{\varepsilon}{2}\langle a\rangle^2\end{aligned} \quad (5.11)$$

Using the cavity output relation (Eq. 3.63), the steady state solution for the output harmonic field is given by

$$B_{\text{out}} = \frac{\sqrt{2\kappa_{\text{out}}^b}\varepsilon\langle a\rangle^2}{2\kappa_b} \quad (5.12)$$

This shows that the output harmonic field is a function of the nonlinear coupling coefficient and the intensity of the circulating fundamental field. The generated second harmonic field is used to drive the Optical Parametric Oscillator, thus the name of 'pump' field.

### 5.3.4 Optical Parametric Oscillator (OPO)

The *Optical Parametric Oscillator* or *OPO*, is the $\chi_2$ nonlinear optical device that generates the squeezed vacuum states.

The Langevin equations for an OPO are given by

$$\dot{a} = -\kappa_a a + \varepsilon a^\dagger b + \sqrt{2\kappa_{in}^a} A_{in} + \sqrt{2\kappa_{out}^a}\delta A_v + \sqrt{2\kappa_l^a}\delta A_l \tag{5.13}$$

$$\dot{b} = -\kappa_b b + \sqrt{2\kappa_{in}^b} B_{in} + \sqrt{2\kappa_{out}^b}\delta B_v + \sqrt{2\kappa_l^b}\delta B_l \tag{5.14}$$

Compared to Eqs. 5.9 and 5.10, the above Equations have a further simplification employed, that is, the term $\frac{\varepsilon}{2}a^2$ in the pump field cavity equation is now zero. This is the assumption that the power in the pump field is not depleted by the parametric process [17], which nullifies the pump field dependence on the fundamental field. This assumption is valid for Optical Parametric Oscillators that operate below threshold ($\varepsilon b < \kappa_a$), which is the true for the case of OPOs producing squeezed vacuum states.

We explore the classical and semiclassical models, whereby the pump field $b$ is treated as a classical field, and combined with the nonlinearity term $\varepsilon$ to create a new term for the total nonlinear gain $g = \varepsilon b$. The classical model will provide a method to obtain measurement of the pump power required to reach threshold, an important experimental parameter. The semiclassical model demonstrates the squeezed state generation behaviour from the OPO.

### 5.3.5 *Classical Behaviour of the OPO*

Considering the classical behaviour of the OPO, the equations for the fundamental field are

$$\dot{a} = -\kappa_a a + g a^\dagger + \sqrt{2\kappa_{in}^a} A_{in} \tag{5.15}$$

$$\dot{a}^\dagger = -\kappa_a a + g^* a + \sqrt{2\kappa_{in}^a} A_{in}^\dagger \tag{5.16}$$

Using the cavity input-output relations, and assuming (without loss of generality) that the amplitude of the input field $A_{in}$ is real, the ratio of the steady-state transmitted output seed power ($P_{\text{out}}$) to the transmitted stready-state output seed power with no total nonlinear gain ($P_{\text{out}}^0$ where $g = 0$) can be expressed as

$$\frac{P_{\text{out}}}{P_{\text{out}}^0} = \frac{(1 + g/\kappa_a)^2}{(1 - |g|^2/\kappa_a^2)^2} \tag{5.17}$$

The term $g/\kappa_a$, also known as the normalised pump parameter $x$ [18], is equivalent to the ratio of the pump field amplitude $B$ to the pump field amplitude at threshold $B_{\text{thr}}$ [8]

$$x = \frac{g}{\kappa_a} = \frac{B}{B_{\text{thr}}} \tag{5.18}$$

thus Eq. 5.17 can be re-written as

$$\frac{P_{out}}{P_{out}^0} = \frac{(1 + B/|B_{thr}|)^2}{(1 - |B/B_{thr}|^2)^2} \tag{5.19}$$

There exist distinct cases for amplication and de-amplification of the output beam, arising from the sign change due to the phase of the pump field. In the case of amplification, the pump field amplitude value is positive ($B > 0$), while for de-amplification the value is negative ($B < 0$). Denoting (for clarity) the output seed power in the amplification case by $P_{out}^+$, in the de-amplification case by $P_{out}^-$, we can equate

$$\frac{B}{B_{thr}} = \frac{\Upsilon_1}{\Upsilon_2} \tag{5.20}$$

where

$$\Upsilon_1 = \sqrt{\frac{P_{out}^+}{P_{out}^-}} - 1, \qquad \Upsilon_2 = \sqrt{\frac{P_{out}^+}{P_{out}^-}} + 1 \tag{5.21}$$

thus Eq. 5.19 can be re-written as

$$\frac{P_{out}^{\pm}}{P_{out}^0} = \frac{[1 + \Upsilon_1/\Upsilon_2]^2}{[1 \mp (\Upsilon_1/\Upsilon_2)^2]^2} \tag{5.22}$$

Figure 5.5 shows a plot of the gain expressions versus input pump power (normalised to threshold power $P_{th} = |B_{thr}|^2$). As the input pump power approaches threshold, the amplification curve asymptotes to $+\infty$. This is a result of the

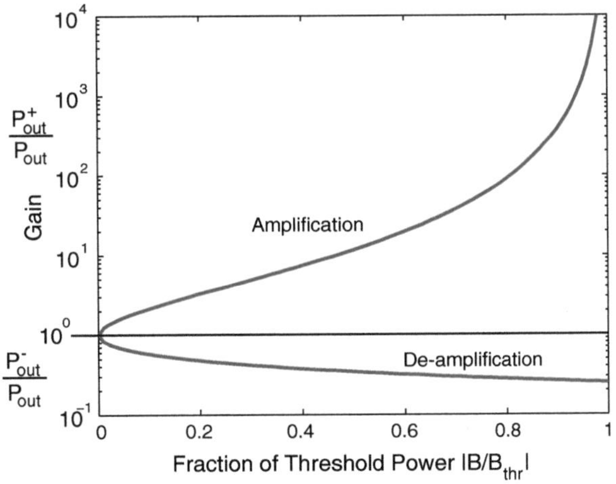


**Fig. 5.5** OPO nonlinear gain curves, showing the behaviour in both amplification and deamplification regimes. As the input pump power approaches threshold, the amplification curve asymptotes to $+\infty$. This is a result of the assumption of no pump depletion breaking down at (and above) threshold powers

assumption of no pump depletion breaking down at (and above) threshold. By injecting a seed field[1] and making measurements of output seed power over several input pump powers lower than threshold, the OPO pump threshold power can be inferred.

### 5.3.6 Semiclassical Behaviour of the OPO

Focussing on squeezing generation, we return to Eq. 5.13, with semiclassical pump ($\varepsilon b = g$). Consider the fluctuating components, given by

$$\delta\dot{a} = (-\kappa_a + g)\delta a + \sqrt{2\kappa_{in}^a}\delta A_{in} + \sqrt{2\kappa_{out}^a}\delta A_v + \sqrt{2\kappa_l^a}\delta A_l \tag{5.23}$$

$$\delta\dot{a}^\dagger = -(\kappa_a + g)\delta a^\dagger + \sqrt{2\kappa_{in}^a}\delta A_{in}^\dagger + \sqrt{2\kappa_{out}^a}\delta A_v^\dagger + \sqrt{2\kappa_l^a}\delta A_l^\dagger \tag{5.24}$$

Calculating the quadratures gives

$$\delta\dot{X}_1^a = (-\kappa_a + g)\delta X_1^a + \sqrt{2\kappa_{in}^a}\delta X_1^{A-in} + \sqrt{2\kappa_{out}^a}\delta X_1^{A-v} + \sqrt{2\kappa_l^a}\delta X_1^{A-l} \tag{5.25}$$

$$\delta\dot{X}_2^a = (-\kappa_a - g)\delta X_2^a + \sqrt{2\kappa_{in}^a}\delta X_2^{A-in} + \sqrt{2\kappa_{out}^a}\delta X_2^{A-v} + \sqrt{2\kappa_l^a}\delta X_2^{A-l} \tag{5.26}$$

Solving in the Fourier domain thus results in

$$\delta\tilde{X}_{1,2}^a = \frac{\sqrt{2\kappa_{in}^a}\delta\tilde{X}_{1,2}^{A-in} + \sqrt{2\kappa_{out}^a}\delta\tilde{X}_{1,2}^{A-v} + \sqrt{2\kappa_l^a}\delta\tilde{X}_{1,2}^{A-l}}{i\omega + \kappa_a \mp g} \tag{5.27}$$

Applying the transmission input-output relation (Eq. 3.63), we find the transmitted noise quadratures

$$\delta\tilde{X}_1^{out} = \frac{(2\kappa_{out}^a - i\omega - \kappa_a + g)\delta\tilde{X}_1^{A-v} + \sqrt{4\kappa_{in}^a\kappa_{out}^a}\delta\tilde{X}_1^{A-in} + \sqrt{4\kappa_l^a\kappa_{out}^a}\delta\tilde{X}_1^{A-l}}{i\omega + \kappa_a - g} \tag{5.28}$$

$$\delta\tilde{X}_2^{out} = \frac{(2\kappa_{out}^a - i\omega - \kappa_a - g)\delta\tilde{X}_2^{A-v} + \sqrt{4\kappa_{in}^a\kappa_{out}^a}\delta\tilde{X}_2^{A-in} + \sqrt{4\kappa_l^a\kappa_{out}^a}\delta\tilde{X}_2^{A-l}}{i\omega + \kappa_a + g} \tag{5.29}$$

which, assuming all input noise variances are vacuum states ($V = 1$), results in noise variances of

$$V_1^{out} = 1 + \frac{\kappa_{out}^a}{\kappa_a}\frac{4(g/\kappa_a)}{\omega^2/\kappa_a^2 + (1 - g/\kappa_a)^2} \tag{5.30}$$

$$V_2^{out} = 1 - \frac{\kappa_{out}^a}{\kappa_a}\frac{4(g/\kappa_a)}{\omega^2/\kappa_a^2 + (1 + g/\kappa_a)^2} \tag{5.31}$$

---

[1] This is the low-power field from Sect. 5.2.

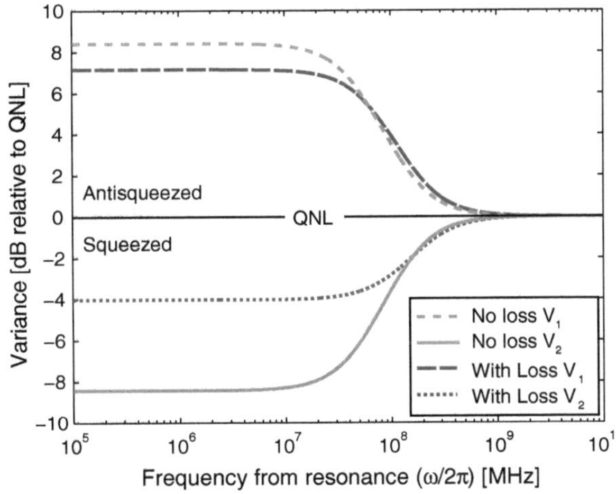


**Fig. 5.6** Output noise variances of an OPO cavity, showing the generation of squeezing and anti-squeezing. Cavity parameters L = 1; $\lambda = 1\ \mu$m; $T_{out} = 0.1$; $x = g/\kappa_a = 0.45$. For the lossless case $T_{in} = T_l = 0$, for the case with non-zero losses $T_{in} = 0.01$, $T_l = 0.001$

Figure 5.6 shows the noise variances as a function of frequency (from cavity resonance). The generated squeezing/anti-squeezing is filtered by the cavity linewidth, thus returning the squeezed state to the vacuum state at higher frequencies. In the lossless case, the magnitude of squeezing and anti-squeezing are equal. With intra-cavity loss, the magnitude of the squeezing/anti-squeezing is lessened, with more pronounced effect in the squeezing quadrature.

#### 5.3.6.1 Escape Efficiency of the OPO

The *escape efficiency*, given by

$$\eta_{esc} = \kappa^a_{out}/\kappa_a \tag{5.32}$$

is the measure of the efficiency that squeezing may exit the OPO cavity. Rewriting Eqs. 5.30 and 5.31 with the escape efficiency and normalised nonlinear gain parameter $x$, gives

$$V_1^{out} = 1 + \eta_{esc}\frac{4x}{\omega^2/\kappa_a^2 + (1-x)^2} \tag{5.33}$$

$$V_2^{out} = 1 - \eta_{esc}\frac{4x}{\omega^2/\kappa_a^2 + (1+x)^2} \tag{5.34}$$

To increase the escape efficiency, one can either:

- Increase the output coupler decay rate $\kappa^a_{out}$ to dominate the total decay rate $\kappa^a$, i.e. $\kappa^a_{out} \gg \kappa^a_{in}, \kappa^a_l \rightarrow \kappa^a \approx \kappa^a_{out}$, or
- Decrease the other contributions to the total decay rate so that the (fixed) output coupler decay rate dominates, i.e. $\kappa^a_{in}, \kappa^a_l \approx 0 \rightarrow \kappa^a \approx \kappa^a_{out}$.

The usual method to increase escape efficiency is to increase the output coupler decay rate (by lowering the output coupling mirror reflectivity). By increasing the output coupler decay rate, the total decay rate becomes larger. However, from Eq. 5.18, an increased total decay rate $\kappa^a$, means that a higher proportion of pump power relative to the threshold pump power is needed to obtain adequate levels of nonlinear gain. Further increasing the escape efficiency is limited by having the necessary pump power available. This is the reason why knowledge of the threshold pump power $P_{\text{th}}$ is needed.

### 5.3.7 The OPO and the Auxiliary Field

The above discussion of the OPO focusses on the second harmonic field and the squeezing fundamental field. The third incident field on the OPO, the Auxiliary Laser field was not discussed. In the OPO quantum Langevin equation above, no terms for the Auxiliary field were included. This is because the presence of the Auxiliary laser field *does not impact on the squeezing generation process*. The Auxiliary and squeezing fundamental fields are mutually exclusive, hence can be treated separately. The Auxiliary field and the Coherent Sideband technique will be discussed in Sect. 6.3.

## 5.4 Detection Stage (Purple Section)

Direct detection of squeezed vacuum states is not possible due to their low optical power signals being masked by technical noise sources, such as thermal noise of electronics. Balanced detection schemes "amplify" up low-power signals to levels above technical noise sources. To measure squeezed vacuum states, and their noise quadratures, balanced detection methods are used.

### 5.4.1 Balanced Detection Schemes

The general schematic for a balanced detection setup is shown in Fig. 5.7. Two fields, the signal field ($A$), and *Local Oscillator (LO)* field ($B$), are mixed on a 50:50 ("balanced") beamsplitter. The relative phase between the two input fields is denoted by the term $\phi = \phi_B - \phi_A$. The resulting fields, $C$ and $D$, are detected by photodetectors at the two output ports of the beamsplitter. Measurements about the properties of the input fields can be made via the electronic subtracting ($i_-$) and addition ($i_+$) of the two photocurrents.

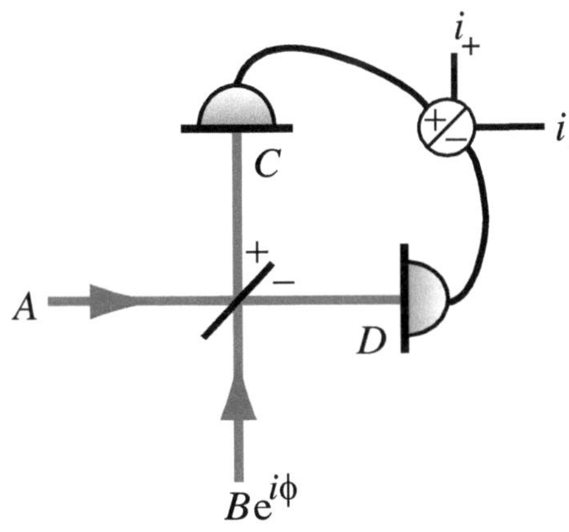


**Fig. 5.7** Schematic of a balanced detection scheme, showing signal ($A$) and LO ($B$) fields, with relative phase $\phi$, and output fields $C$, $D$. The quantities of interest are contained in the difference and sum photocurrents $i_-$, $i_+$. Reprinted with permission from [19], copyright (2014) by IOP Publishing

Different detection schemes arise from this general setup through the different choices of local oscillator field:

- If the Local Oscillator is a bright coherent field, but not at the same wavelength as the signal field, then the detection scheme is known as *balanced heterodyne detection*. This form of detection scheme is used in telecommunications, spectroscopy and astronomy [20–22].
- If the Local Oscillator is a bright coherent field at same wavelength as the signal field, then the detection system is called a *balanced homodyne detector*. Applications for balanced homodyne detectors include measurements in Quantum Cryptography and Magnetic Resonant Imaging [23, 24].

The choice of detection scheme depends on the characteristics of the signal field of interest. We focus on the balanced homodyne detector, known for remainder of this thesis as the homodyne detector. This method is routinely used in detecting quadrature noise properties of signal fields, thus making it ideal for detecting quadrature squeezed states.

### *5.4.2 Balanced Homodyne Detection (HD)*

Utilising the beamsplitter convention (Sect. 3.4.2) and linearisation formalism (Sect. 3.4.1), the output fields of the homodyne detector ($C$, $D$) are:

$$C = \frac{1}{\sqrt{2}}\left((\bar{A} + \delta A) + (\bar{B} + \delta B)e^{i\phi}\right) \tag{5.35}$$

$$D = \frac{1}{\sqrt{2}}\left((\bar{A} + \delta A) - (\bar{B} + \delta B)e^{i\phi}\right) \tag{5.36}$$

The difference photocurrent between the two detectors is therefore given by

$$\begin{aligned} i_- &= D^\dagger D - C^\dagger C \\ &= (\bar{A} + \delta A)^\dagger (\bar{B} + \delta B) e^{i\phi} + (\bar{B} + \delta B)^\dagger e^{-i\phi} (\bar{A} + \delta A) \end{aligned} \tag{5.37}$$

To simplify, we note that:

- Cross-fluctuation terms, e.g. $\delta A \delta B$, are negligible
- Mean field amplitudes are assumed to be real, i.e. $\bar{A} = \bar{A}^*$, $\bar{B} = \bar{B}^*$
- The signal mean amplitude is much less than the LO mean amplitude, i.e. $\bar{A} \ll \bar{B}$. This is known as the *LO condition*, resulting in terms not containing $\bar{B}$ to be neglected. This assumption holds for squeezed vacuum signals.

Using the quadrature definitions, the difference photocurrent is now

$$i_- = 2\bar{A}\bar{B}\cos\phi + \bar{B}(\delta X_1^A \cos\phi + \delta X_2^A \sin\phi) \tag{5.38}$$

The difference current contains information about the signal field amplitude ($\delta X_1^A$) and phase ($\delta X_2^A$) quadratures, "amplified" by the LO mean field amplitude ($\bar{B}$). The LO amplification raises the quadrature effects of the signal over the technical noise levels in the photodetectors. Therefore, the difference photocurrent of balanced homodyne detection can facilitate the measurement of squeezed vacuum quadratures. The choice of quadrature (or combination of quadratures) observed is determined by the choice of the relative phase $\phi$.

The sum photocurrent is given by

$$i_+ = \bar{A}^2 + \bar{A}\delta X_1^A + \bar{B}^2 + \bar{B}\delta X_1^B \tag{5.39}$$

$$\approx \bar{B}^2 + \bar{B}\delta X_1^B \tag{5.40}$$

This is equivalent to a direct measurement of the LO field.

## 5.5 Filtering and Stabilisation Stages

This section describes the two subsystems that are use for spatial mode filtering and pump power stabilisation.

### *5.5.1 Mode Cleaner (MC)*

The *mode cleaner* is an optical device used to prepare the Local Oscillator beam for the homodyne detector, so that it is spatially filtered to a known transverse profile, and provide a reference origin-point for the beam. This omponent is necessary for low-frequency detection of squeezed states. The role of the mode cleaner for low-frequency detection of squeezed states is expanded further in Chap. 7—*Properties affecting Squeezing Magnitude and Audio-frequency Measurement*. Optical cavities and single-mode optical fibres can be used as mode cleaners.

### 5.5.2 Mach-Zehnder Interferometer (MZ)

The *Mach-Zehnder* interferometer is used for pump power stabilisation in the squeezer via the stabilisation of transmitted pump power by controlling the relative interferometer arm lengths. The Mach-Zehnder was first utilised for pump power stabilisation in a squeezer by Khalaidovski et al. [25], while other interferometer types had been used in other experiments for power stabilisation (for example in [26]).

## 5.6 Control Stages

Driving an experiment parameter to a reference value, and maintaining that parameter at that reference value (within a specified accuracy) is often needed in experiments. An active feedback control system measures its own output (of a particular parameter, such as transmitted optical power) and uses that measurement to feed back and drive the system to the desired state.

Figure 5.8 shows a generic control loop diagram of an active feedback control system. The system component whose output that we wish to control is called the Plant, and the Sensor measures the output of the Plant. The Sensor's output is compared to the reference value with the Comparator. The difference between the Sensor output and the reference is called the *error signal*. The error signal is sent to the Controller, which changes the Plant output via the Actuator. If the Sensor measurement and reference value are approximately equal, the error signal value is small and thus the Plant output is as required [27].

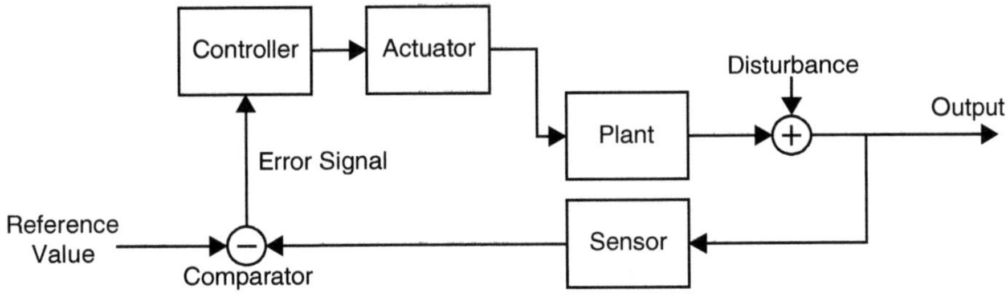


**Fig. 5.8** Block diagram of a generic active feedback control loop

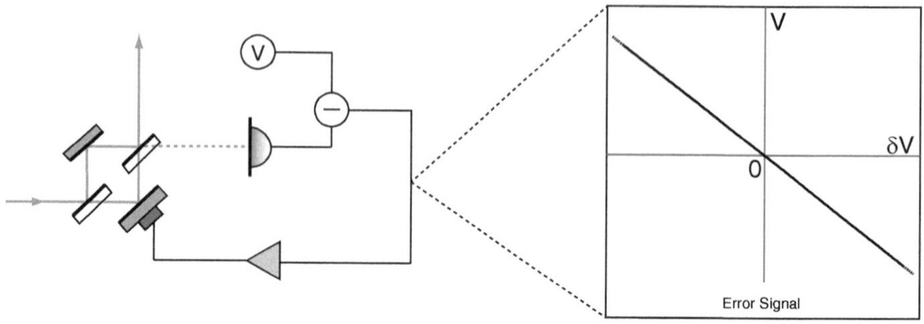


**Fig. 5.9** An example loop diagram and error signal for the DC voltage subtraction locking technique. The loop actuates on the Mach-Zehnder such that photodetector output voltage (from incident optical power) matches the reference voltage

The squeezer uses four feedback control loop scheme-types. Three feedback control loop types [Loops (1), (2) and (3) of Fig. 5.1] are now briefly introduced. The fourth, the Coherent Sideband locking technique will be explored in Chap. 6 (Sect. 6.3).

### 5.6.1 DC Voltage Subtraction Locking

Figure 5.9 shows an example loop diagram and error signal for the DC voltage subtraction locking technique. This techique is used for controlling the Mach-Zehnder output, hence the Plant is a Mach-Zehnder in Fig. 5.9. The error signal is derived by taking the difference ($\delta V$) between the sensor (photodiode) output voltage and a desired reference voltage. The control loop then drives the Mach-Zehnder until the sensor output voltage matches the desired reference voltage ($\delta V = 0$). This technique has the advantage of being relatively simple to implement. However, the lock stability of the control loop depends on the stability of the reference voltage.

### 5.6.2 Offset Frequency Locking

Offset frequency locking is used to set a fixed frequency relationship between two lasers. This technique is used to maintain the offset between the Main and Auxiliary lasers' optical frequency. Shown in Fig. 5.10, the readout is the optical beat note between the two laser fields, whose beat note frequency is the difference in optical frequency between the two laser sources ($f$). By comparing the beatnote to a reference sinusoid with the desired frequency offset ($f_0$), this forms the error signal. When the error signal is zero, the beatnote frequency and desired frequency must be equal. When the error signal is not zero, the frequency of one of the lasers needs to be tuned in order to return to the desired offset. This is enabled by actuating on the laser's PZT or temperature.

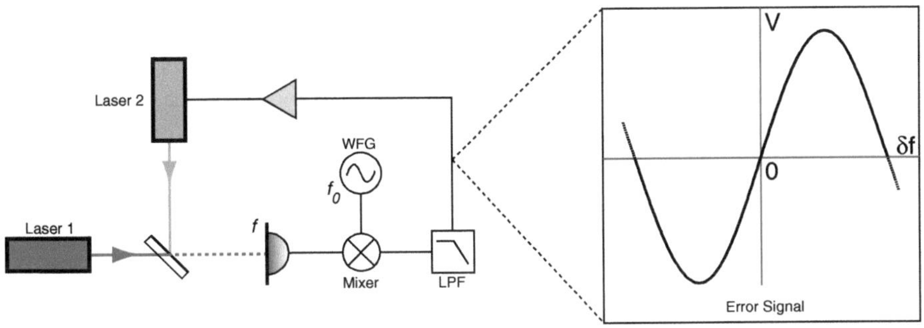


**Fig. 5.10** Example schematic and error signal for the offset frequency locking technique. The optical beat note between two lasers is compared to the reference sinusoid at the desired offset frequency $f_0$ from the waveform generation (WFG), thus forming a sinusoidal error signal [with locking points mod($2\pi$)]. After low-pass filtering, the frequency of one of the lasers is actuated to maintain the optical beat note at the desired frequency, hence controlling a fixed-frequency offset between the two lasers

### *5.6.3 Pound-Drever-Hall Locking*

Pound-Drever-Hall locking, or *PDH* locking, is a commonly used technique in optical experiments that require stabilisation of (I) the path length of optical cavities to maintain resonance with an incident laser field [28], or (II) the optical frequency of an incident laser field to the resonance of a stable optical reference cavity [29, 30]. Case (I) is presented here, used to control the optical path length of the Mode Cleaner, SHG and OPO cavities. For a more detailed treatment of the PDH locking technique, [30] is a recommended source.

Shown in Fig. 5.11, phase modulation sidebands are placed on the incident carrier field with a phase modulator (PM), and the path length change of the cavity is encoded with the relative phase shift between the sidebands and carrier field.

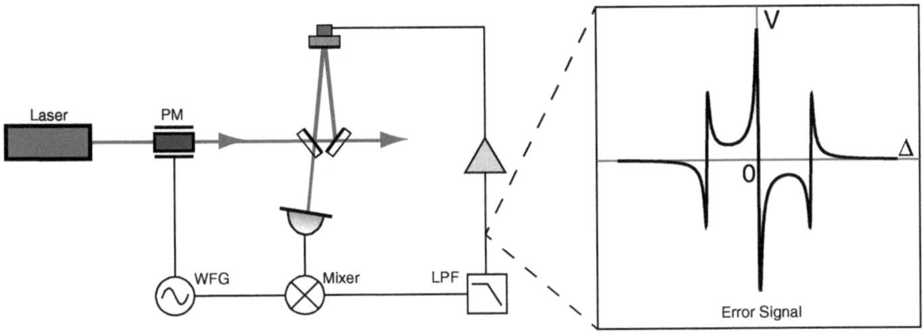


**Fig. 5.11** Example schematic and error signal for the Pound-Drever-Hall locking technique. The schematic is detailed in Sect. 5.6.3. The cavity resonance point is located at the origin. The central locking point in distinguished from the other zero-crossings by their opposite-signed (negative/positive) slopes

After signal processing, an error signal is produced that is zero on resonance and approximately linear around the resonance point. Therefore the error signal is proportional to displacement of the cavity path length from resonance, or the cavity detuning $\Delta$. This is then fed back to the cavity length actuator to correct the displacement, returning the cavity to resonance. On cavity resonance, the detuning is zero ($\Delta = 0$). This validates the simplifying assumption of $\Delta = 0$ used in the quantum Langevin equations.

## 5.7 Summary

An overview of a squeezed state generator or squeezer was presented, canvassing the:

- Laser input stages—The sources of laser light for the apparatus.
- Nonlinear optical stages—The Second Harmonic Generator (SHG), and the crucial Optical Parametric Oscillator that generates the squeezed states.
- Detection stage—Balanced homodyne detection that enables measurement of noise quadratures.
- Filtering and stabilisation stages—Mach-Zehnder and Mode Cleaner devices
- Control loops—DC voltage subtraction locking, Offset frequency locking and Pound-Drever-Hall locking techniques.

## References


1. C. Gross, T. Zibold, E. Nicklas, J. Estève, M.K. Oberthaler, Nonlinear atom interferometer surpasses classical precision limit. Nature **464**, 1165–1169 (2010)
2. C.D. Hamley, C.S. Gerving, T.M. Hoang, E.M. Bookjans, M.S. Chapman, Spin-nematic squeezed vacuum in a quantum gas. Nat. Phys. **8**, 305–308 (2012)
3. N. Korolkova, G. Leuchs, R. Loudon, T.C. Ralph, C. Silberhorn, Polarization squeezing and continuous-variable polarization entanglement. Phys. Rev. A **65**, 052306 (2002)
4. J.F. Corney, J. Heersink, R. Dong, V. Josse, P.D. Drummond, G. Leuchs, U.L. Andersen, Simulations and experiments on polarisation squeezing in optical fibre (2008)
5. T. Corbitt, Y. Chen, F. Khalili, D.Ottaway, S. Vyatchanin, S.Whitcomb, et al., Squeezed-state source using radiation-pressure-induced rigidity. Phys. Rev. A **73**, 023801 (2006)
6. F. Marino, F.S. Cataliotti, A. Farsi, M.S. de Cumis, F. Marin, Classical signature of ponderomotive squeezing in a suspended mirror resonator. Phys. Rev. Lett. (2010)
7. D.W.C. Brooks, T. Botter, S. Schreppler, T.P. Purdy, N. Brahms, D.M. Stamper-Kurn, Non-classical light generated by quantum-noise-driven cavity optomechanics. Nature **488**, 476–480 (2012)
8. B.C. Buchler, Electro-optic control of quantum measurements. Ph.D. thesis, Physics Department, Australian National University, Canberra, Australia, 2008
9. K. McKenzie, Squeezing in the audio gravitational wave detection band. Ph.D. thesis, Physics Department, Australian National University, Canberra, Australia, 2008
10. H. Vahlbruch, Squeezed light for gravitational wave astronomy. Ph.D. thesis, Albert Einstein Institute and Leibniz University of Hannover, Hannover, Germany, 2008

11. T. Eberle,V. Händchen, J. Duhme, T. Franz, R.F. Werner, R. Schnabel, Strong Einstein-Podolsky-Rosen entanglement from a single squeezed light source. Phys. Rev. A **83**, 052329 (2011)
12. N. Lee, H. Benichi, Y. Takeno, S. Takeda, J. Webb, E. Huntington, A. Furusawa, Teleportation of nonclassical wave packets of light. Science **332** (2011)
13. R. Boyd, *Nonlinear Optics* (Academic Press, San Diego, 1992)
14. B.E.A. Saleh, M.C. Teich, *Fundamentals of Photonics* (Wiley, New York, 1991)
15. D.S. Hum, M.M. Fejer, Quasi-phasematching. C.R. Phys. **8**, 180–198 (2007)
16. M.L. Bortz, M.A. Arbore, M.M. Fejer, Quasi-phase-matched optical parametric amplification and oscillation in perodically poled LiNbO3 waveguides. Opt. Lett. **20**(1) (1995)
17. D.F. Walls, G. Milburn, *Quantum Optics*, 2nd edn. (Springer, Berlin, 2008)
18. T. Aoki, G. Takahashi, A. Furusawa, Squeezing at 946 nm with periodically poled $KTiOPO_4$. Opt. Express **14**, 6930–6935 (2006)
19. S.S.Y. Chua, B.J.J. Slagmolen, D.A. Shaddock, D.E. McClelland, Quantum squeezed light in gravitationalwave detectors. Class. Quantum Gravity **31**, 183001 (2014)
20. L.G. Kazovsky, N.S. Kopeika, Heterodyne detection through rain, snow, and turbid media: affective receiver size at optical through millimeter wavelengths. Appl. Opt. **22**(5) (1983)
21. A.L. Betz, M.A. Johnson, R.A. McLaren, E.C. Sutton, Heterodyne detection of CO2 emission lines and wind velocities in the atmosphere of Venus. Astrophys. J. **208**(2), L141 (1976)
22. M.D. Levenson, B.A. Paldus, R.N. Zare, Optical heterodyne detection for cavity ring-down spectroscopy. US Patent, 6094267 (2000)
23. T. Hirano, H. Yamanaka, M. Ashikaga, T. Konishi, R. Namiki, Quantum cryptography using pulsed homodyne detection. Phys. Rev. A **68** (2003)
24. D.C. Noll, D.G. Nishimura, Homodyne detection in magnetic resonance imaging. IEEE Trans. Med. Imaging **10**(2) (1991)
25. A. Khalaidovski, H. Vahlbruch, N.Lastzka, C. Gräf, K. Danzmann, H. Grote, et al., Long-term stable squeezed vacuum state of light for gravitational wave detectors. Class. Quantum Gravity **29**, 075001 (2012)
26. C.M. Mow-Lowry, B.S. Sheard, M.B. Gray, D.E. McClelland, S.E. Whitcomb, Experimental demonstration of a classical analog to quantum noise cancellation for use in gravitational wave detection. Phys. Rev. Lett. **92**, 161102 (2004)
27. A. Abramovici, J. Chapsky, *Feedback Control Systems—A Fast-Track Guide for Scientists and Engineers* (Kluwer Academic Publishers, Boston, 2000)
28. M. Rakhmanov, R.L. Savage Jr., D.H. Reitze, D.B. Tanner, Dynamic resonance of light in fabry perot cavities. Phys. Rev. A **305**, 239–244 (2002)
29. R.W.P. Drever, J.L. Hall, F.V. Kowalski, J. Hough, G.M. Ford, A.J. Munley, H. Ward, Laser phase and frequency stabilization using an optical resonator. Appl. Phys. B **31**(2), 97–105 (1983)
30. E.D. Black, An introduction to Pound-Drever-Hall laser frequency stabilization. Am. J. Phys. **69**, 79 (2001)

# Part II
# The Doubly Resonant, Travelling-Wave Squeezed Light Source

# Chapter 6
# The Doubly Resonant, Travelling-Wave Squeezed Light Source

**Abstract** Squeezed states applicable for enhancing gravitational-wave detectors is the driver behind the development of large squeezing magnitudes at audio-detection band frequencies. This chapter describes the squeezed light source that achieved a greater than 10 dB squeezing measurement across the audio gravitational-wave detection band (10 Hz–10 kHz). This squeezing level was generated using a doubly resonant, travelling-wave optical parametric oscillator (OPO) with a wedged non-linear crystal, and controlled using a modified coherent sideband locking technique. This chapter presents the development and results of the doubly resonant, travelling-wave squeezed light source, grouped into four topics:

1. The features of the squeezed light source OPO and modified coherent sideband locking scheme are first presented. Sections 6.1 and 6.2 present the design choices and parameters of the OPO. Section 6.3 introduces the coherent sideband locking technique, followed by the changes implemented for the modified coherent sideband locking technique.
2. Various experiment parameters and properties that impact on the squeezing measurement are then presented. These are categorised into properties affecting squeezing magnitude measurement (Sect. 6.4), and properties affecting low frequency squeezing measurement (Sect. 6.5).
3. Upgrades of the squeezed light source resulting from the understanding of the above properties are then detailed in Sect. 6.6. The upgraded squeezed light source configuration is presented in Sect. 6.7.
4. Lastly, the results of measurements from the squeezer are presented and characterised in Sect. 6.8. This includes the first measurement of greater than 10 dB squeezing across the audio gravitational-wave detection band (11.6±0.4 dB above 200 Hz), and 5900 s of continuously controlled squeezing.


S.S.Y. Chua, *Quantum Enhancement of a 4 km Laser Interferometer Gravitational-Wave Detector*, Springer Theses,
DOI 10.1007/978-3-319-17686-4_6

## 6.1 Design Properties of the Optical Parametric Oscillator

Squeezed states have been generated from optical parametric oscillators with differing cavity configurations and optical properties [1–3]. The reasoning for the design choices of the doubly resonant, travelling-wave OPO are presented in this section. This configuration was also used for the LIGO Squeezed Light Injection Experiment (Part 3 of the thesis).

### *6.1.1 Bow-Tie Travelling-Wave Cavity*

Optical cavities can be broadly divided into two categories: travelling-wave and standing-wave configurations [4]. If the round-trip intra-cavity beam passes any given point only in one propagation direction, the configuration is a travelling-wave cavity. Complement to this, if the round-trip intra-cavity beam passes a given point in two propagation directions, the configuration is a standing-wave. Figure 6.1 shows example cavities of these two configuration categories.

The travelling-wave configuration was chosen as the geometry for the OPO. The first reason is that a travelling-wave configuration is first-order immune to backscattered light, due to the fact that backscattered light propagating in the reverse direction will couple into the reverse intra-cavity mode. The removal of backscattered light is a particularly important factor for low frequency squeezing. Backscattered light and the travelling-wave configuration is explored in depth in Chap. 7. The second reason is the availability of multiple spatially-separated input/output ports for optical beams, used for OPO operation and characterisation, to access the cavity.

There are drawbacks with using a travelling-wave configuration. Compared with a standing-wave cavity, there are an increased number of optical components. This reduces total mechanical stability, as well as increase optical losses due to the increased number of surfaces. Beam astigmatism is also an issue, as intra-cavity fields are no longer incident on optics at normal-incidence [4]. The 'bow-tie' configuration has minimal beam astigmatism for a travelling-wave geometry, as all beam incident angles within the cavity are small, typically around 6°–8° from normal incidence.

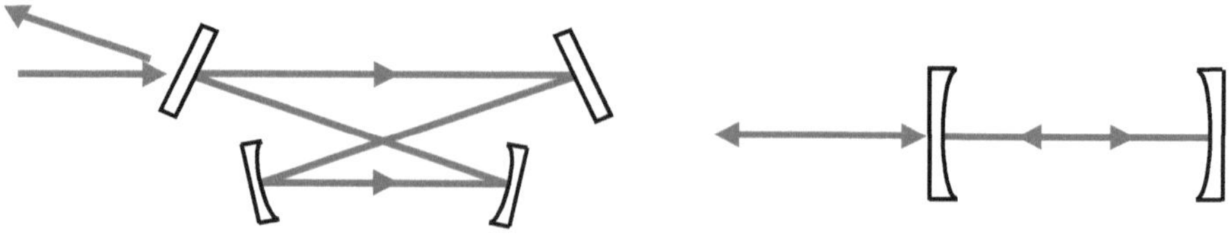

**Fig. 6.1** Optical cavity configurations. *Left* Travelling-wave 'bow-tie' cavity. *Right* Standing-wave 'Fabry-Perot' cavity

### 6.1.2 Doubly Resonant Cavity

In the context of this thesis, a *doubly resonant* OPO cavity represents a cavity that is simultaneously resonant at pump and fundamental frequencies. This is opposed to a singly resonant OPO system that is only resonant at the fundamental (squeezing) frequency. The design of the OPO is a doubly resonant cavity.
The advantages of a doubly resonant OPO are:

- The ease of obtaining a cavity-length control signal, via the use of the pump field as the readout field.
- The assurance of perfect mode matching of the interacting fields for the nonlinear process. Since the pump and fundamental fields share the same optical cavity, the ratio of the beam waist sizes is $\sqrt{2}$, which is the ratio that satisfies the Boyd-Kleinman condition for optimal second-order nonlinear interaction [5].
- Photothermal effects induced by higher-order spatial modes of the pump beam are minimised, as those spatial modes not matched to the cavity mode will be filtered when the cavity is on resonance.
- The pump field amplitude is resonantly enhanced, thus for a given input pump power a higher nonlinear gain can be reached. This allows the possibility of increasing the transmission of the output coupler at the fundamental frequency to increase the escape efficiency ($\eta_{esc}$).

There are also disadvantages associated with a doubly resonant OPO that require management. These are:

- The photothermal effect associated with absorption of the pump field (in the nonlinear crystal) is greater due to the resonantly-enhanced pump field. This can cause length instabilities, such as optical bi-stability [6]. Management of the crystal temperature can minimise such effects.
- The intra-cavity dispersion of the fundamental and harmonic fields causes the resonance frequencies of the two fields to be offset with each other. When co-resonance of both fields does not occur, the nonlinear interaction is suppressed [7]. To compensate for the wavelength dispersion, a wedged nonlinear crystal is used.

### 6.1.3 Wedged Nonlinear Media

Dispersion compensation in doubly resonant nonlinear systems has been previously achieve by the insertion of a pellicle plate [8, 9]. However, the presence of a compensation plate introduces two additional optical surfaces within the optical cavity, increasing optical losses. In lieu, dispersion compensation can be achieved by modifying the nonlinear crystal geometry such that an angled-wedge is polished for one of the surfaces [10, 11].

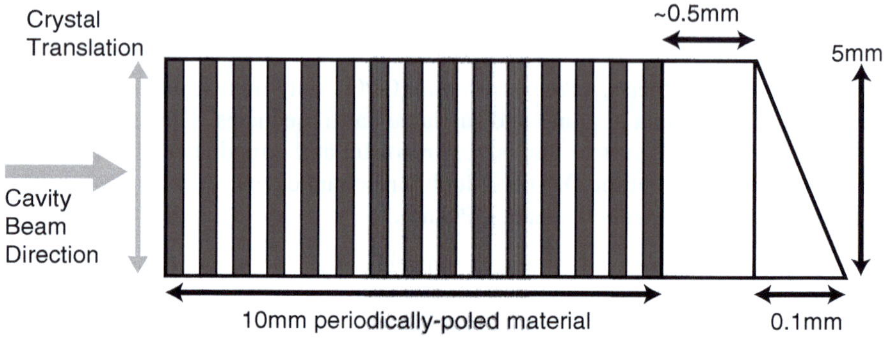


**Fig. 6.2** Illustration of the OPO nonlinear crystal of PPKTP, with angle-polished wedge in non-poled KTP material. Poling domains are not scale

Figure 6.2 illustrates such a wedged crystal. By lateral translation of the crystal position perpendicular to the optical beam propagation direction, the effective path length is changed, due to the refractive index of the additional optical material. Thus, a crystal position can be found where dispersion compensation can be achieved. Dual resonance for the nonlinear process, along with no introduction of additional optical surfaces, is realised simultaneously.

## 6.2 The Doubly Resonant, Travelling-Wave OPO Cavity

The layout of the doubly resonant, travelling-wave OPO cavity is shown in Fig. 6.3. The cavity design is based on the OPO cavity used in work completed by Grosse [12], with modifications to improve cavity portability and ease of cavity construction. The OPO cavity parameters are summarised in Table 6.1.

The OPO cavity consists of four half-inch mirrors with reflective coatings at 532 and 1064 nm to form a doubly resonant cavity in travelling-wave, bow-tie

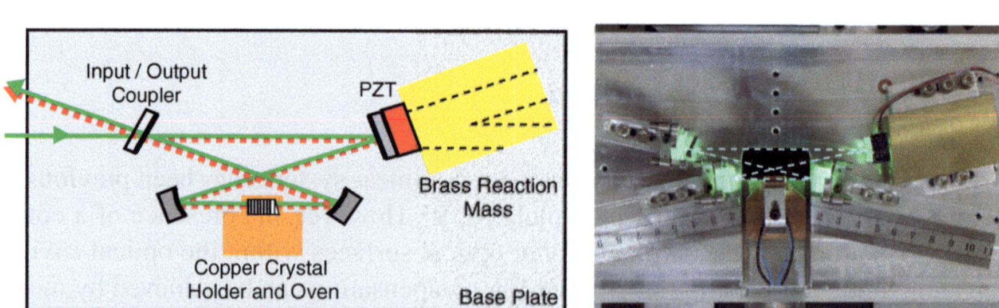


**Fig. 6.3** Schematic and photo of the doubly resonant, travelling-wave OPO Cavity. The bow-tie optical path has been highlighted in the photo

**Table 6.1** Optical and cavity parameters of the DB optical parametric oscillator

| Cavity parameter | Symbol | Value | Units |
|---|---|---|---|
| Fundamental (squeezing) wavelength | $\lambda_a$ | 1064 | nm |
| Second harmonic (pump) wavelength | $\lambda_b$ | 532 | nm |
| Round-trip optical path length | L | 0.279 | m |
| Curved mirrors' radius of curvature | RoC | −38 | mm |
| Input/output coupler reflectivity (fundamental) | $R^a_{in/out}$ | 0.839 | – |
| Input/output coupler reflectivity (pump) | $R^b_{in/out}$ | 0.722 | – |
| PZT mirror reflectivity | $R^a_2$ | 0.9982 | – |
| Total intra-cavity loss at fundamental | $T^a_l$ | 0.0026 | – |
| Total intra-cavity loss at pump | $T^b_l$ | 0.046 | – |
| Finesse (fundamental) | $\mathscr{F}_a$ | 35.3 | – |
| Finesse (pump) | $\mathscr{F}_b$ | 16.8 | – |
| Full-width-half-maximum (fundamental) | $\Delta\nu_a$ | 31.2 | MHz |
| Full-width-half-maximum (pump) | $\Delta\nu_b$ | 65.3 | MHz |
| Escape efficiency | $\eta_{esc}$ | $98.5 \pm 0.3$ | – |
| Threshold power | $P_{thr}$ | $108 \pm 5$ | mW |

configuration. The reflective coatings of the mirrors were manufactured for an angle of incidence of 6°, resulting in 12° cavity-vertex angles. The cavity-vertex angles provided the smallest angle of incidence possible for minimising astigmatism without intra-cavity beam clipping from cavity mirror mounts. The round-trip optical path length was approximately 28 cm, so that the cavity benchtop footprint was minimised given the cavity-angle constraints. The input/output coupler reflectivities for the two wavelengths were chosen to optimise squeezing escape efficiency given the available pump power (see Sect. 5.3.6). The second flat mirror is mounted to a ring piezo-electric transducer (PZT) on a brass reaction-mass block. The brass block had recessed hollows to allow access for optical beams to the second flat mirror.

The nonlinear medium of the OPO used is a quasi-phase-matched (Sect. 5.3.1) PPKTP crystal, anti-reflection coated at both wavelengths, with a wedge of

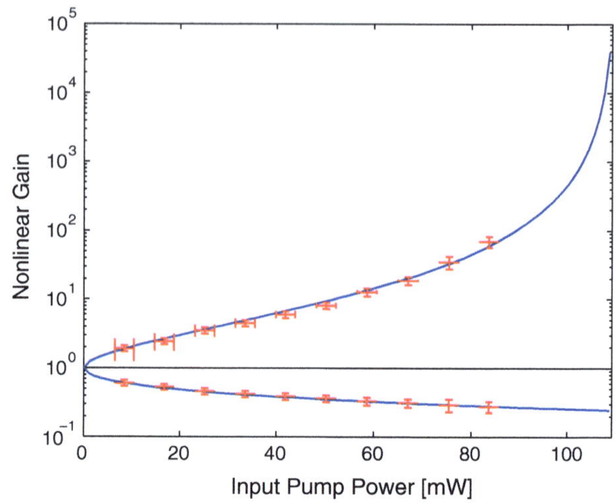


**Fig. 6.4** The DB-OPO cavity—nonlinear gain as a function of input pump power. The fitted threshold pump power is 108 ± 5 mW

angle-polished non-poled material (Sect. 6.1.3). PPKTP was chosen for its high nonlinear gain coefficient and low phase matching temperature [13]. The crystal is housed in a temperature-controlled peltier oven, mounted on a five-axis alignment stage to allow crystal position translation. This translation, in combination with the crystal wedge, enables the compensation of dispersion. The entire cavity is housed on a single aluminium block, allowing for easy construction and mobility of the OPO cavity as a complete unit.

A measurement of the OPO nonlinear gain as a function of input pump power was made, with the data shown in Fig. 6.4. The pump threshold power (Sect. 5.3.5), was thus fitted to be 108 ± 5 mW. For the remainder of this thesis, the OPO will be referred to as the *doubly-resonant bow-tie OPO or DB-OPO* at ANU.

## 6.3 Modified Coherent Sideband Locking

Squeezed state generation presents only part of the challenge for applying squeezed states for gravitational-wave detector enhancement. In addition to the control loops of Sect. 5.6, the squeezing ellipse phase needs to be controlled relative to the measurement quadrature. Various techniques for controlling the squeezing ellipse phase have been developed [1, 14], with the technique currently used for squeezed-vacuum phase control called Coherent Sideband Locking. This technique is used for the GEO squeezed-light source [15], as well as the LIGO H1 Squeezed Light Injection Experiment.

This section first introduces the Coherent Sideband locking technique. This is followed by details of a modified technique based on Coherent Sideband locking to control the squeezing ellipse phase.

### 6.3.1 Coherent Sideband Locking

The Coherent Sideband locking technique was proposed by Chelkowski et. al [16] and experimentally demonstrated by Vahlbruch et al. [17]. The simplified schematic and phasor diagrams are shown in Fig. 6.5. For this technique, a second bright optical field of amplitude $\alpha_\Omega$ at an offset frequency ($\Omega$) from the fundamental frequency ($\omega_c$), is injected into the OPO [Point (a)]. This *coherent sideband* field undergoes the same nonlinear interaction as the squeezed state generation process, being amplified or deamplified as per the pump phase at the crystal. This nonlinear process can be viewed as the creation of a second sideband of the offset bright field, and described as (Eq. 9 of [16])

$$\begin{aligned} E^{CL} &\propto \frac{1+g}{\sqrt{2g}}\alpha_\Omega \cos(\omega_0 t + \Omega t) - \frac{1-g}{\sqrt{2g}}\alpha_\Omega \cos(\omega_c t - \Omega t - 2\phi) \\ &\propto \cos(\omega_c t + \Omega t) - A\cos(\omega_c t - \Omega t - 2\phi) \end{aligned} \tag{6.1}$$

where $\sqrt{g} = \exp(R)$ with $R$ the squeezed factor (Sect. 3.2.3), $A$ is the relative amplitude of the two coherent locking sidebands,[1] and $\phi$ is the relative phase of the pump and coherent sideband fields.

The leakage and reflected fields are then measured [Point (b) – Loop 4a], to be demodulated at $2\Omega$ to obtain an error signal (Eq. 12 of [16])

$$S^{CL} \propto \frac{(-1+g^2)\alpha_\Omega^2}{4g}\sin(2\phi) \tag{6.2}$$

which can be used to lock the relative phase pump field and coherent sideband field ($\phi \to 0$). The sideband field also co-propagates with the squeezed state towards the homodyne detector [Point (c)]. The detected field on the homodyne detector [Point (d) – Loop 4b] is used to generate an error signal with the LO field, via demodulation at frequency $\Omega$ (Eq. 17 of [16])

$$S^{LO} \propto \frac{\sqrt{2}\alpha^{LO}\alpha_\Omega^2(-1+g)}{\sqrt{g}}\sin(2\phi + \Phi) \tag{6.3}$$

which in combination with the first coherent sideband locking loop, can then control the relative phase $\Phi$ of the coherent sideband field and the LO field. The two demodulating signal generators are electronically phase locked together, thus the squeezing ellipse is controlled.

The implementation of the coherent sideband locking scheme involves firstly generating the offset sideband field. Initial demonstration of the technique [17] saw the use of an acousto-optic modulator (AOM) to create the field, via upshifting of a bright seed field. However, due to residual seed field being present from imperfect

[1] From Eq. 9 of [16], $A$ is calculated as $A = \frac{1-g}{1+g}$.

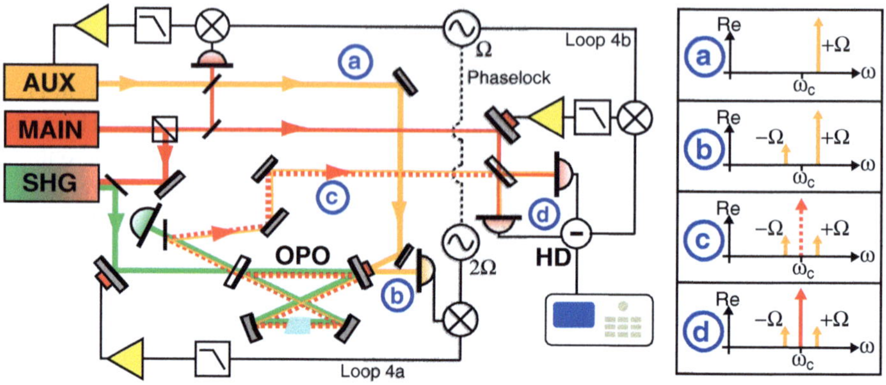


**Fig. 6.5** Coherent sideband locking schematic, and phasor diagrams corresponding to specific optical field points within the schematic. Mach-Zehnder interferometer and mode cleaner are removed from the schematic for clarity. Details discussed in Sect. 6.3.1. Loop numbering (*2*, *4a* and *4b*) reflect the same numbering as Fig. 5.1 of Chap. 5

AOM modulation, current implementations of the technique use a second phase-locked laser—the Auxiliary laser (AUX). The second implementation step involves the addition and modification of photodetectors. The photodetector at the reflected coherent port [Loop 4a in Fig. 6.5] needs to be AC coupled and have a detection response resonance at the needed $2\Omega$ frequency. This is to ensure a decent signal-to-noise ratio. The homodyne detector response also needs to be modified in order for a measurement at the offset frequency $\Omega$ for the second lock [Loop 4b].

Compared to other squeezing ellipse phase control methods, the first advantage of this technique is the removal of bright fields at the fundamental (squeezing) frequency, a concern with squeezed vacuum state detection. The second advantage is this technique allows the tuning of the squeezing ellipse phase to any angle as needed, via the choice of relative phase between the two demodulating signal generators.

### *6.3.2 Modified Coherent Sideband Locking*

A modified coherent sideband locking scheme was developed and implemented for the DB-OPO squeezed light source. The modified coherent sideband locking setup and phasor diagram is shown in Fig. 6.6.

- The modified scheme firstly involves the single-pass frequency-doubling of the Auxiliary laser beam to form a co-propagating auxiliary second harmonic with auxiliary field [Point (A) in Fig. 6.6]. This was done using a standalone SHG (SHG2). After attenuation of the auxiliary field, these two beams are now interfered with the pump beam on a dichroic beamsplitter (High reflection HR at 1064 nm, anti-reflection AR at 532 nm) (DBS) in the pump field path.

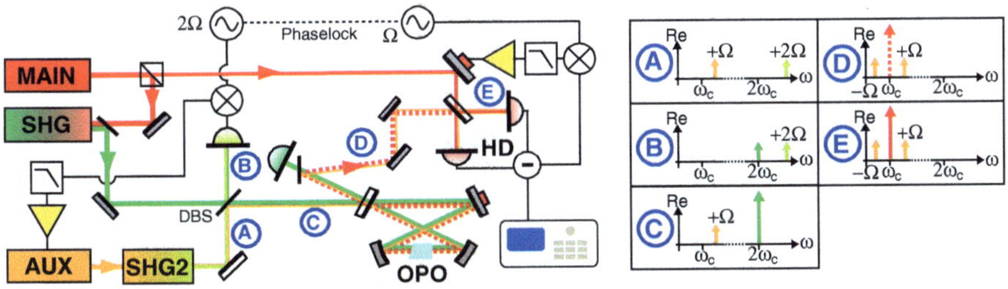


**Fig. 6.6** Modified coherent sideband locking schematic, and phasor diagrams corresponding to specific optical field points within the schematic. Details discussed within Sect. 6.3.2. Reprinted with permission from [18], copyright (2011) by the Optical Society of America

- The interference and beat note of the two green fields at $2\Omega$ is now detected [Point (B)], which is then fed back to the auxiliary laser to maintain the Main-Auxiliary laser frequency offset at $\Omega$. This also sets the phase relationship between the reflected auxiliary field and pump field inherently, from the dichroic beamsplitter point, as the optical paths are now common for the two beams from that point.
- The auxiliary field reflected off the DBS now enters the OPO cavity via the input coupler [Point (C)], not a high-reflectivity mirror, allowing more auxiliary field to sample the nonlinear interaction within the OPO. The auxiliary field (with the second sideband) then co-propagates with the squeezed field [Point (D)] to the homodyne detector, where the second lock is the same as with the unmodified coherent sideband locking scheme [Point (E)].

A comparison between original and modified coherent sideband locking techniques is summarised in Table 6.2. The advantages of the modified method are the inherent setting of the phase relationship between auxiliary field and pump field that removes an entire control loop, and the increased loop control bandwidths. A comparison of loop bandwidths is shown in Fig. 6.7. The auxiliary-pump phase reference is now incorporated with the Main-Auxiliary laser offset frequency loop, leveraging the Auxiliary Laser PZT actuation range. An increase of unity-gain bandwidth from 200 Hz (trace b) to ~40 kHz (trace d) was obtained.

The auxiliary field now couples in via the input coupler, instead of a high-reflectivity mirror. Having a greater coupling to the nonlinear interaction enables a larger second sideband amplitude to be generated. Also, the promptly reflected field increases the coherent signal beat note detected by the homodyne detector. Thus, the larger signal increases the optical gain of the second coherent locking loop, resulting in a the unity-gain bandwidth increase from 45 Hz to ~25 kHz (trace e). Larger magnitude and higher frequency disturbances can now be tracked, making a more robust loop with higher dynamic range. The modified coherent sideband scheme improves squeezing measurement as the squeezing ellipse angle is better controlled. The impact of squeezing ellipse angle is discussed further in Sect. 6.4.4.

The disadvantages are the requirement of a second SHG, and the extra care needed with the promptly reflected auxiliary laser field. The promptly reflected field can

**Table 6.2** Comparison between original and modified coherent sideband locking schemes

| Original scheme | Modified scheme |
|---|---|
| Three control loops | (A) Two control loops |
| | (D) Additional passive SHG required |
| (A) Spatial filtering of the auxiliary field with the OPO | (D) Auxiliary field prompt-reflection co-propagating with the squeezed field |
| (D) Less auxiliary field interacting with nonlinear process | (A) More field interacting with nonlinear process—increased signal strength |
| (D) Smaller control loop bandwidths | (A) Larger loop control bandwidths |
| (D) More auxiliary power incident on the OPO required | (A) Less required auxiliary power incident on the OPO |

Advantages and disadvantages are denoted with (A) and (D) respectively

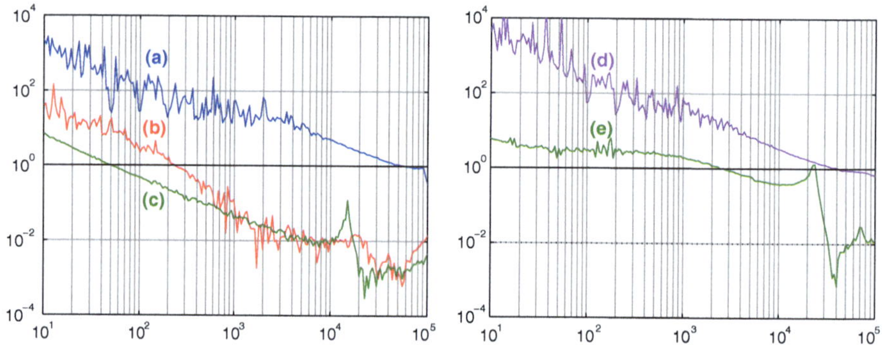


**Fig. 6.7** Coherent sideband locking control loop bandwidths. *Left* Original scheme with 140 μW of coherent beam power incident on the OPO, with traces **a** main and auxiliary laser loop, **b** sideband-pump phase loop [4a] and **c** sideband-LO phase loop [4b]. *Right* Modified scheme with <10 μW of coherent beam power incident on the OPO, with traces **d** main and auxiliary laser + pump phase reference loop and **e** sideband-LO phase loop

potentially impact on the squeezing measurement, for example, by introducing additional noise that can overlap and pollute the squeezing measurement frequencies. Choosing the auxiliary laser frequency offset so that a shot-noise limited operating regime overlaps with the squeezing measurement frequencies, and using low auxiliary laser optical power (yet keeping a reasonable loop bandwidth) minimises the potential impact.

## 6.4 Properties Affecting Squeezing Magnitude Measurement

Observation of squeezing magnitude provides certainty in the squeezing being generated and available for quantum-noise enhancement. Various properties can adversely affect the measurement of squeezing magnitude. These are optical loss, homodyne spatial mode mismatch and squeezing ellipse phase noise. These properties are presented in this section, as an understanding of their impact was crucial to measuring greater than 10 dB squeezing.

### 6.4.1 Optical Losses

Similarly to the photodetector with losses (Sect. 3.4.3), the presence of loss on an optical field ($A$) can be interpreted as the introduction of a partially transmissive mirror, with power transmission $\eta_l$. Of particular importance to squeezing measurement, this "loss mirror" also couples in vacuum fluctuations ($\delta v$). The field and photon number of a beam with optical loss ($A'$) is given by

$$A' = \sqrt{\eta_l}A + \sqrt{1-\eta_l}\delta v \tag{6.4}$$

$$A'^{\dagger}A' = \left(\eta_l(|A|^2 + A\delta X_1^a) + \sqrt{\eta_l(1-\eta_l)}A\delta X_1^v\right) \tag{6.5}$$

The variance of the photon number relative to quantum noise limit can be calculated as

$$\begin{aligned} V_{A'}^{out} &= \frac{\eta_l|A|^2(\eta_l V_A^{in} + (1-\eta)V_A^v)}{\eta_l|A|^2} \\ &= \eta_l V_A^{in} + (1-\eta)V_A^v \\ &= \eta_l V_A^{in} + (1-\eta_l) \end{aligned} \tag{6.6}$$

with the variance of the vacuum field being unity. This illustrates that *presence of optical loss returns the noise variance of the optical field back towards the quantum noise limit.*

It follows that the impact of optical losses can greatly hinder the amount of squeezing measurable. Figure 6.8 illustrates the effect of optical loss on squeezing magnitude measurement for different loss percentages. It also illustrates the larger attenuation impact with greater input squeezing. For example, with 50 % optical loss, 5 dB generated squeezing becomes 1.8 dB measured squeezing, a reduction of 3.2 dB squeezing. This is compared with 40 dB generated squeezing becoming 3 dB measured squeezing, an attenuation of 37 dB.

Optical loss enters the measurement through imperfect optical components in the squeezing propagation path between the OPO and the balanced homodyne, as well as the imperfect detection efficiency of the measurement photodiodes. These are represented as efficiencies $\eta_{prop}$ and $\eta_{pd}$ respectively.

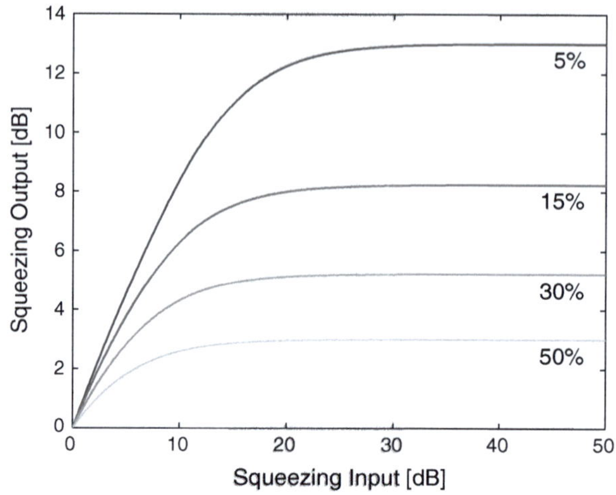


**Fig. 6.8** Optical loss impact on measured squeezing magnitude, as a function of input squeezing magnitude. Optical loss curves of 5, 15, 30 and 50 % are shown. Reprinted with permission from [19], copyright (2014) by IOP Publishing

## *6.4.2 Spatial Mode Mismatch*

Spatial mode mismatch between the squeezed field and the homodyne Local Oscillator field also degrades squeezing measurement. This is viewed as the introduction of higher-order modes into the system, which are occupied by the unsqueezed vacuum state. Consider the homodyne setup in Fig. 6.9a. The signal and local oscillator fields are spatially matched with efficiency $\eta_h$, with the remaining Local Oscillator field interacting with an orthogonal higher-order spatial mode occupied by the vacuum field $\delta v^H$.

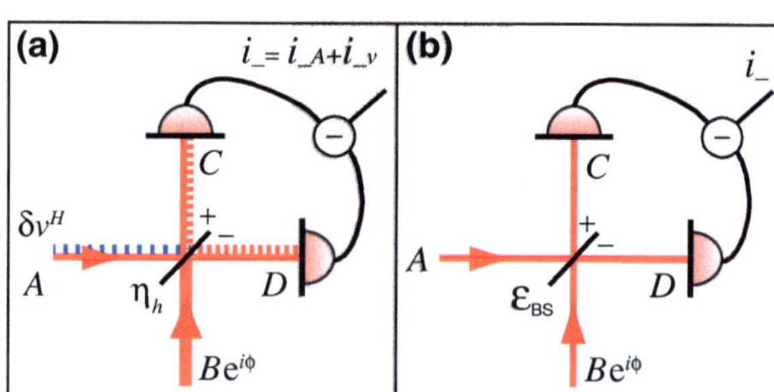


**Fig. 6.9** Non-perfect balanced homodyne detection. **a** Homodyne with spatial mode mismatch, showing the mixing in of the (higher-order mode) vacuum state $\delta v^h$ (Sect. 6.4.2). **b** Homodyne with unbalanced arm powers, represented by splitting ratio $\varepsilon_{BS} \neq 1/2$ (Sect. 6.5.2)

Similar to the calculation presented for the perfect homodyne detector in Sect. 5.4.2, we can write the difference photocurrents for each of the signals $(A,\ \delta v^H)$. This gives

$$i_{-A} = 2\bar{A}\sqrt{\eta_h}\bar{B}\cos\phi + \sqrt{\eta_h}\bar{B}(\delta X_1^A\cos\phi + \delta X_2^A\sin\phi) \tag{6.7}$$

$$i_{-v} = \sqrt{1-\eta_h}\bar{B}(\delta X_1^{vH}\cos\phi + \delta X_2^{vH}\sin\phi) \tag{6.8}$$

The total difference photocurrent of the system is then the sum of Eqs. 6.7 and 6.8

$$\begin{aligned} i_- &= 2(\sqrt{\eta_h}\bar{A})\bar{B}\cos\phi \\ &+\bar{B}\left((\sqrt{\eta_h}\delta X_1^A + \sqrt{1-\eta_h}\delta X_1^{vH})\cos\phi\right. \\ &\left.+(\sqrt{\eta_h}\delta X_2^A + \sqrt{1-\eta_h}\delta X_1^{vH})\sin\phi\right) \end{aligned} \tag{6.9}$$

Comparing Eq. 6.9 to Eq. 5.38 for the perfect balanced homodyne, the consequence of spatial mode mismatch is that it is *equivalent to introducing another optical loss term on to the signal field, with efficiency* $\eta_h$. Therefore the spatial mode matching of the squeezing and Local Oscillator fields is important for effective squeezing measurement.

To assess the spatial mode matching, the *fringe visibility* $\mathscr{V}$ is measured, given by

$$\mathscr{V} = \frac{P_{\text{max}} - P_{\text{min}}}{P_{\text{max}} + P_{\text{min}}} \tag{6.10}$$

with $P_{\text{max}}$, $P_{\text{min}}$ the maximum and minimum power of interference fringes incident on one of the homodyne photodetectors. Explicitly, these maximum and minimum powers are

$$P_{\text{max}} = \frac{1}{2}(|A|^2 + \eta_h|B|^2 + 2\sqrt{\eta_h}|AB| + (1-\eta_h)|B|^2) \tag{6.11}$$

$$P_{\text{min}} = \frac{1}{2}(|A|^2 + \eta_h|B|^2 - 2\sqrt{\eta_h}|AB| + (1-\eta_h)|B|^2) \tag{6.12}$$

Using Eqs. 6.11 and 6.12, calculating the fringe visibility with equal beam optical powers (i.e. $|A|^2 = |B|^2 = |AB|$), results in

$$\begin{aligned} \mathscr{V} &= \frac{2\sqrt{\eta_h}|AB|}{|A|^2 + \eta_h|B|^2 + (1-\eta_h)|B|^2} \\ &= \sqrt{\eta_h} \end{aligned} \tag{6.13}$$

Thus the homodyne efficiency can be measured with the fringe visibility, $\eta_h = \mathscr{V}^2$.

### 6.4.3 *Squeezing Measurement Efficiency*

To characterise the overall impact of optical loss on squeezing magnitude measurement, the optical loss of the device is the combination of the optical-loss sources introduced above. The *detection efficiency* is the multiplicative combination of the propagation, homodyne and photodetection efficiencies, or

$$\eta_{det} = \eta_{prop} \times \eta_h \times \eta_{pd} \tag{6.14}$$

The *total measurement efficiency* is the product of the detection efficiency with the OPO escape efficiency.

$$\eta_{tot} = \eta_{esc} \times \eta_{det} \tag{6.15}$$

This leads to the updating of Eqs. 5.33 and 5.34 for the quadrature variances of the OPO output with optical loss present, given by

$$V_{1,2}^{out} = 1 \pm \eta_{tot} \frac{4(x)}{\omega^2/\kappa_a^2 + (1 \mp x)^2} \tag{6.16}$$

### 6.4.4 *Squeezing Ellipse Phase Noise*

Squeezing ellipse phase noise refers to the residual fluctuations of the squeezing ellipse when being controlled to the measurement quadrature of interest. This effect on squeezing measurement has been studied extensively [1, 20–22]. As shown in Fig. 6.10, fluctuations of the squeezing ellipse couple noise from the anti-squeezed quadrature into the squeezing measurement, thus a lower squeezing magnitude is

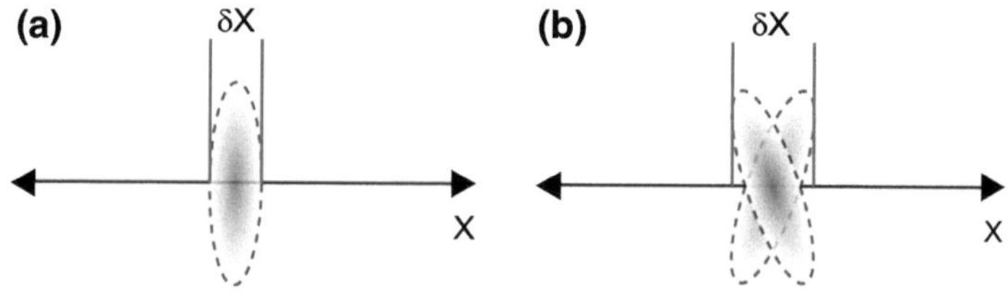


**Fig. 6.10** The effect of squeezing ellipse phase noise on the vacuum squeezing measurement. **a** No ellipse phase noise, thus the measurement quadrature (X) measures the true squeezing magnitude ($V_X = \langle \delta X \rangle^2$). **b** Ellipse phase noise present, thus the squeezing ellipse is pivoting. This couples in the antisqueezed noise quadrature. The result is the measurement quadrature no longer reflects, and is diminished compared to, the true squeezing magnitude

measured than the true squeezed quadrature. For root-mean-square (rms) phase fluctuations $\tilde{\theta}$, the change in the measured quadrature variance is given by [1]

$$V_{1,2}^{out} \approx V_{1,2}\cos^2\tilde{\theta} + V_{2,1}\sin^2\tilde{\theta} \tag{6.17}$$

### 6.4.5 *Squeezing/Antisqueezing as a Function of OPO Nonlinear Parametric Gain*

Equations 6.16 and 6.17 describe the impact of squeezing measurement efficiency and squeezing phase noise on squeezed state measurement. Measured values of these parameters are important for characterising the performance of a squeezed-light source. In experiment, these parameters can be obtained or verified with a measurement of the *squeezing and antisqueezing levels as a function of OPO nonlinear parametric gain.*

The output of such a measurement for three different rms phase noise values ($\tilde{\theta} = 0$, 20 and 40 mrad) is illustrated in Fig. 6.11. The antisqueezing quadrature is very weakly affected by squeezing ellipse phase noise, thus the antisqueezing data can be used to infer the total measurement efficiency. With the total measurement efficiency inferred, the squeezing data can be used to determine the squeezing ellipse phase noise. Figure 6.11 shows the distinct variations between the curves of different squeezing ellipse phase noise levels.

To obtain such a measurement, multiple squeezing and anti-squeezing traces are taken as the OPO operating gain point is varied via the increase or decrease of incident pump power. Once squeezing ellipse phase noise and measurement efficiency are determined, the detection efficiency can also be obtained using knowledge of

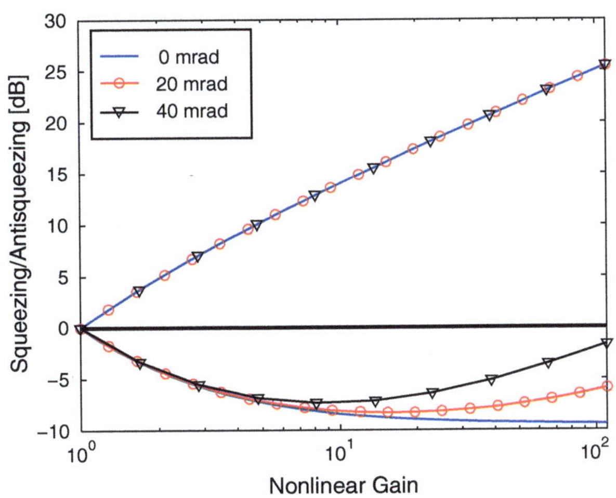


**Fig. 6.11** Squeezing and antisqueezing magnitude, as a function OPO nonlinear parametric gain with different squeezing rms phase fluctuations, $\tilde{\theta} = 0$, 20 and 40 mrad. This illustration is generated using a total efficiency of $\eta_{tot} = 0.89$. The antisqueezed data can be used to constrain the total measurement efficiency, while the squeezing data can constrain the squeezing ellipse phase noise

the OPO escape efficiency (Eq. 5.32 with measurements of the OPO cavity mirror reflectivities). Results of such squeezing/antisqueezing quadrature measurement runs will be used to characterise the DB-OPO and results of the LIGO Squeezed Light Injection Experiment.

## 6.5 Properties Affecting Squeezed State Detection at Low Frequencies

The first measurements of squeezing and shot noise in the audio frequency band were limited by a 'roll-up' in noise at low frequencies [23]. From these observations, an open question arose—was the roll-up in noise caused by degradation in squeezed state generation or in squeezing measurement. In subsequent work, it was determined that the degradation was from issues with squeezing measurement [24]. Published work brought together the many issues faced with low frequency measurement [25], and steps to their mitigation. This section describes the properties that affect the measurement of squeezed light at low frequencies.

### *6.5.1 Detection Electronics*

The electronic components used in photodetectors impact the measurement performance at low frequencies. Flicker noise is an electronic noise source found in resistors [26]. The level of flicker noise is proportional to the amount of electric current passing through the resistor, and results in a 1/f spectrum roll-up towards low frequencies. Flicker noise was determined to be the main cause of the 'roll up' in the first squeezing measurements.

The source of flicker noise is not well known, but can be reduced using wire-wound or metal-film-type resistors [27]. By replacing the resistor components with low flicker noise metal-electrode-leadless-face (or MELF) type resistors, flat shot-noise homodyne measurements down to 5 Hz were achieved.

Dark noise is the electronic noise of the detection system when no light is incident on the photodiodes. Particularly at lower frequencies, the dark noise level may dominate the shot noise and squeezing noise levels, limiting their measurement. Careful selection of electronic components ensures that the dark noise level is well below the shot noise and squeezing levels. The *dark noise clearance level* (*that the shot noise level is above the dark noise level*) of off-the-shelf photodetectors are approximately 10 dB, and custom-designed photodetectors, such as the homodyne detectors used in experiments within this thesis, can have a dark-noise clearance of at least 20 dB.

### 6.5.2 Homodyne Detection Balancing

Perfect balanced homodyne detection has been previously described in Sect. 5.4.2. However, the degree of balancing, or the signal ratio between the two arms of the homodyne detector, is crucial to the performance of the homodyne detector. This is illustrated by considering the homodyne detector setup shown in Fig. 6.9b, where the reflectivity of the homodyne beamsplitter is now $\varepsilon_{\mathrm{BS}}$. Similar to the derivation found in Sect. 5.4.2, the fields of the two arms $C$, $D$ are given by

$$C = \sqrt{\varepsilon_{\mathrm{BS}}}(A + \delta A) - \sqrt{1 - \varepsilon_{\mathrm{BS}}}(B + \delta B)e^{i\phi} \tag{6.18}$$

$$D = \sqrt{\varepsilon_{\mathrm{BS}}}(B + \delta B)e^{i\phi} + \sqrt{1 - \varepsilon_{\mathrm{BS}}}(A + \delta A) \tag{6.19}$$

resulting in a difference current $i_-$ of

$$\begin{aligned} i_- = {} & \underbrace{(2\varepsilon_{\mathrm{BS}} - 1)(B^2 + B\delta X_1^B)}_{(i)} + \underbrace{(1 - 2\varepsilon_{\mathrm{BS}})(A^2 + A\delta X_1^A)}_{(ii)} \\ & \underbrace{+ 2\sqrt{\varepsilon_{\mathrm{BS}}(1 - \varepsilon_{\mathrm{BS}})}\left[2AB\cos\phi + B(\delta X_1^A \cos\phi + \delta X_2^A \sin\phi)\right]}_{(iii)} \end{aligned} \tag{6.20}$$

The three terms of Eq. 6.20 represent (i) the coupling of the LO field and LO noise into the measurement, (ii) the coupling of the direct signal field and signal noise into the measurement, and (iii) the LO amplification of the signal noise quadratures. Ideally, the splitting between the arms should be equal (i.e. $\varepsilon_{\mathrm{BS}} = \frac{1}{2}$), resulting in the terms (i) and (ii) being zero and term (iii) returning to Eq. 5.38.

Relative intensity noise of the Local Oscillator field can be greater than 30 dB above the shot noise level at low frequencies. Further, the signal field also carries intensity noise from the coherent locking sidebands. These are represented by the fluctuating terms of (i) and (ii) in Eq. 6.20. Should the balancing of the homodyne be inadequate to suppress the intensity noise below the squeezing noise level, then these noise sources will pollute the measurement of squeezing.

The measure of the balancing is the *common-mode rejection* (*CMR*) ratio, where a high CMR represents a well-balanced homodyne detector. The CMR for both the LO port and the signal port need to be maximised. This is initially set with the power splitting of the Local Oscillator and signal beams between the homodyne arms. However, the CMR can be disturbed over time by beam pointing and non-stationary noise. In more detail:

- **Photodiode surface inhomogeneities and Beam pointing**—The quantum efficiency of any photodiode can vary across the photodiode active surface [27, 28]. If the position of an incident beam on the photodiode changes, then via these inhomogeneous photodiode responses, the photocurrent output will vary. *Beam pointing fluctuations* are changes in the trajectory of a beam that result in incident-beam position changes. These pointing fluctuations originate from motion of optical

components, such as low frequency acoustic vibrations and air currents. The combination of beam pointing and photodiode inhomogeneities leads to fluctuations in the CMR and the reentering of the low frequency noise.
Beam pointing can be considered to be the fractional coupling of the original beam mode into different spatial modes. From this perspective, the effect of beam pointing can be greatly filtered by a mode cleaner cavity. Changes in beam pointing are now converted into beam intensity fluctuations, which the balanced homodyne will common-mode reject. The homodyne is still susceptible to beam fluctuations after the mode cleaner. Beam pointing can be reduced here with short beam paths, mechanically-stable optical mounts and minimising air currents.

- **Non-stationary noise**—The mechanism for non-stationary noise is the presence of traversing dust particulates on an optical beam [29]. This affects the CMR in two ways. Firstly, there is a short-live change in the beam's spatial profile, changing the beam mode into a higher-order spatial mode. This change couples similarly as beam pointing and photodiode surface inhomogeneity to changing the CMR, and similarly solved by a mode cleaner cavity. Secondly, traversing dust after the homodyne beamsplitter can cause short-lived but large changes in the balancing due to optical power loss between the two homodyne arms. Minimising air currents and/ or shielding the homodyne arms will reduce the non-stationary noise impact after the homodyne beamsplitter.

### 6.5.3 Backscattered Light Interferences

When light is scattered from an optical beam, and then recombined with the original beam after reflection off an external object (such as another optical element), a *backscattered light interference* or *parasitic interference* is formed [24]. Optical path length changes of the scattered beam, typically caused by displacement motion of the backscattering object, will impart intensity modulations that can impact low frequency noise measurement. In particular, light backscattered from components in the squeezing signal path become spurious signals that are amplified by the homodyne Local Oscillator. Figure 6.12 shows a shot noise measurement with such backscattered light signals polluting the low frequency spectrum measurement.

Possible backscatter objects in the squeezing path include optical mount surfaces, scattering centres such as dust, as well as the direct back-reflection of the OPO cavity itself. The backscatter tolerance of the OPO cavity is the subject of the next Chapter. Mitigation of backscattered noise can be achieved through carefully placed beam blocks stopping the backscattered reflection, or by reducing the displacement motion of the scattering object.

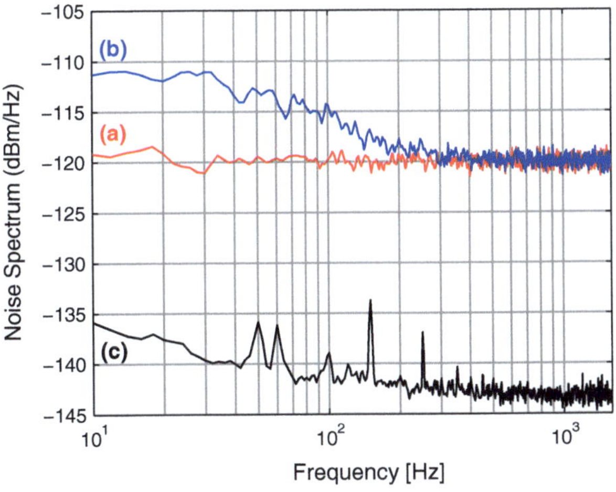


**Fig. 6.12** Backscattered light interferences in low frequency measurement. **a** Measurement of shot noise, with the squeezing-signal path blocked. **b** Measurement of shot noise, with squeezing-signal path open. All normal operating beams on the OPO (pump beam and auxiliary beam) were blocked. Light from the homodyne detector is being scattered up and backscattered along the squeezing-signal beam path, causing backscattered light interferences at low frequencies. **c** Electronic dark noise. Reprinted with permission from [25], copyright (2012) by IOP Publishing

## 6.6 Upgrades to the Squeezed Light Source

Over two and a half years, the squeezed light source underwent a progression of upgrades that reflected developments of understanding about the properties affecting squeezing magnitude and low-frequency squeezed light measurement. This section presents the progression, with the squeezing spectra measured shown in Fig. 6.13.

- **March 2009**—The early stage of the squeezed light source at ANU. The implementation of the unmodified coherent sideband locking (Sect. 6.3.1) was completed, together with a DB-OPO of the design in Sect. 6.1. This was in preparation for the commissioning work of the OPO to be used in the LIGO H1 Squeezed Light Injection Experiment. Squeezing was measured to be 4.4 dB above 2 kHz. The small squeezing magnitude was a result of intra-cavity loss, traced to being from losses of the PPKTP crystal anti-reflection coating ($\kappa_l = 1.4\,\%$). Low-frequency flicker noise from the installed homodyne electronics accounted the roll-up seen below 2 kHz.
- **May 2009**—Considerable effort was undertaken in improving the stability of the control loops, particularly the phase lock between Main and Auxiliary lasers. This improved the squeezing magnitude through reduction of squeezing ellipse phase noise (Sect. 6.4.4). Electronic cabling was improved, reducing in the electric AC-mains frequencies (and harmonics) coupling into measurements. An improvement in the CMR of the signal port was achieved, as it was observed that noise coupled

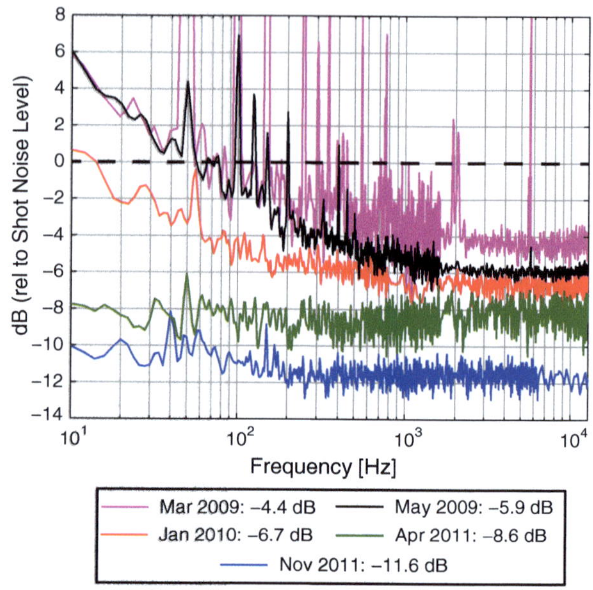


**Fig. 6.13** Improvement of the measured squeezed light source output. Dates correspond to descriptions within Sect. 6.6. November 2011 data measured over three frequency spans: 0–1.6 kHz (RBW = 2 Hz), 0–6.4 kHz (RBW = 8 Hz) and 0–102.4 kHz (RBW = 128 Hz). Remaining traces measured over two frequency spans: 0–1.6 kHz (RBW = 2 Hz) and 0–12.8 kHz (RBW = 16 Hz). All traces have at least 50 rms averages. Dark noise has not been subtracted from any of the data

at low frequencies via the coherent sideband locking field. A reduction of coherent sideband power, from ~2 mW to 140 μW was also needed to reduce the coupling to a level where low frequency measurement was no longer impacted. The homodyne performance was still limited by electronic flicker noise at low frequencies. The commissioning and testing of a LIGO H1 Optical Parametric Oscillator began at this point in time.

- **January 2010**—The OPO for the LIGO H1 Experiment was constructed with a new PPKTP crystal. These new crystals were processed by three companies: Raicol Crystals Ltd. [30] provided the base crystal pieces, Photon LaserOptik GmbH [31] undertook the optical surface and wedge polishing (with a micro-roughness < 0.2 Å rms), and LaserOptik GmbH [32] applied the anti-reflection coatings at the two wavelengths (reflectivity < 0.1 % at 1064 nm and <0.5 % at 532 nm). When installed, the intra-cavity loss of the OPO was measured to be $\kappa_l = 0.26\,\%$, a significant reduction from the 1.4 % intra-cavity loss from the previous crystals. The LIGO H1 Squeezer homodyne detector, designed and supplied by Henning Vahlbruch from the Max Planck Institute for Gravitational Physics [33], arrived as part of the LIGO H1 OPO testing process. This was incorporated into the setup in-lieu of the previous homodyne detector.
  Specialist optical components from ATFilms Inc [34] also arrived at this time. These included high-reflectivity mirrors, dichroic mirrors, and optical lenses, all with super-polishing (micro-roughness < 1 Å rms) and coatings. These replaced all optics in the squeezing signal path. Further a super-polished 50:50 beamsplitter was installed for the balanced homodyne detector.
  A separate experimental setup characterised sources of the low-frequency noise in

homodyne detection [25]. Electronic flicker noise, beam pointing, dust particulates and backscattered light interferences were found to be the major contributors to low-frequency noise. Following on from that work, an isolation tank around the detection components and the mode cleaner cavity in the LO path was introduced. The result was that the shot noise spectrum was measured flat down to below 10 Hz and across the audio detection band. However, the low-frequency roll up in the squeezing spectrum still remained, later determined to be from backscattered-light interferences.

- **April 2011**—With the delivery of the LIGO H1 OPO and LIGO H1 Squeezer homodyne detector for the LIGO Squeezed Light Injection Experiment, the squeezer apparatus underwent another major overhaul of the optical layout. A replacement DB-OPO was constructed, containing one of the new-batch PPKTP crystals (and non-superpolished cavity mirrors), as well as a replacement balanced homodyne detector based on the LIGO H1 Squeezer Homodyne. The optical layout of the OPO and homodyne minimised the optical beam path between these key components. The modified coherent locking technique was implemented in this period, delivering the high range, robust lock of the squeezing ellipse angle. Squeezing measurements were recommenced in October 2011.
  Backscattered light interferences along the squeezing signal path were suspected and later verified to be a major contributor to the low-frequency roll up in the spectrum. Careful hunting of scattering locations and mitigation using IR-absorbing glass was undertaken. It was also suspected that beam pointing may be playing a role in degradation of squeezing, beyond the bounds of the isolation tank. Thus, a full-table enclosure was built around the squeezer. This box helped reduce squeezing ellipse phase noise, beam pointing fluctuations, and protected the entire experiment from further dust particulates. With the full table enclosure, this saw the measurement 8.6 dB squeezing across the audio detection band.
- **November 2011**—The homodyne photodiodes were replaced with photodiodes with high quantum efficiency (quoted as $\eta_{pd} > 99\,\%$). These new photodiodes were measured to have a quantum efficiency approximately 4 % greater than the previous photodiodes. Once installed, the squeezing magnitude measurement reached greater than 10 dB across the audio-band.

## 6.7 The DB-OPO Squeezed Light Source

The schematic of the November 2011 squeezer apparatus is shown in Fig. 6.14. The main laser (MAIN) is a continuous wave Nd:YAG, non-planar ring oscillator (NPRO) [35] laser with operating wavelength at 1064 nm (1.2 W Mephisto—Innolight GmbH [36]). The output light is split into two beams. The first beam, and the majority of the output laser power is sent to the second harmonic generator (SHG) for conversion to 532 nm pump light. The second beam, after spatial mode-filtering by the mode cleaner cavity (MCC) is sent to the balanced homodyne detector (HD) to become the Local Oscillator field.

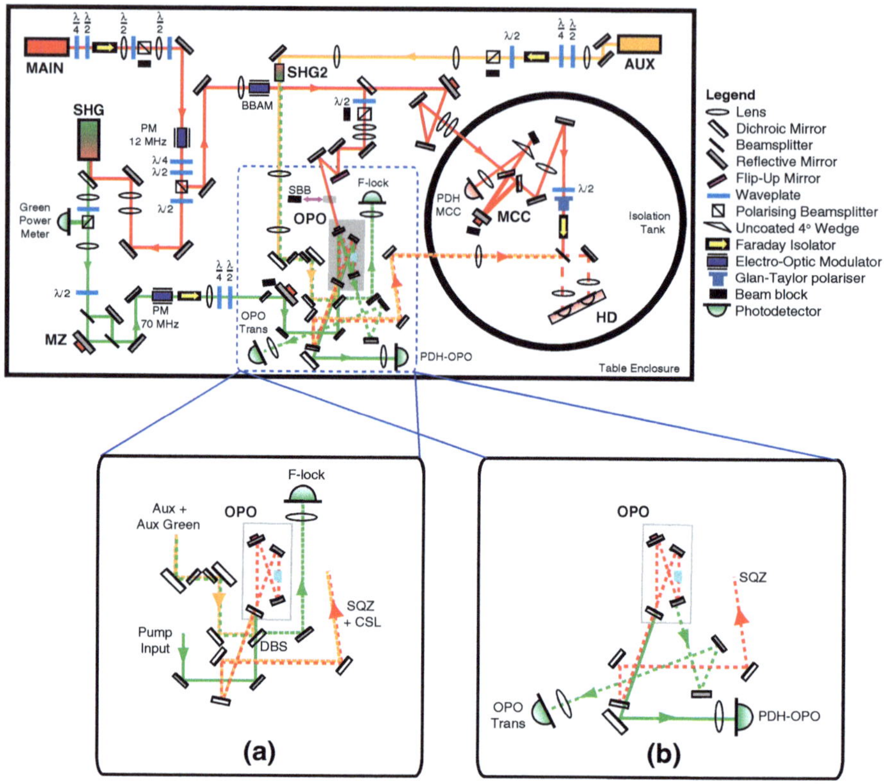


**Fig. 6.14** Schematic of the DB-OPO squeezed light source. The control loop electronics and optical lens values were omitted for clarity. During squeezing measurements, the bright OPO seed field is blocked by the seed beam block (SBB). *BBAM* Broadband electro-optic Amplitude Modulator, *PM* electro-optic Phase Modulator. The *lower panels* expand on the input/output beams of the OPO: **a** Pump and modified coherent locking inputs/outputs, the 'F-lock' green photodetector used to detect the beat note between the pump and green-auxiliary fields at the Dichroic Beamsplitter DBS. **b** OPO pump reflection port used for OPO PDH locking, and OPO pump transmission port used for the DC-voltage subtraction locking of the Mach-Zehnder. *SQZ* Squeezed light, *CSL* Coherent sideband locking field, *Aux* Auxiliary field, *Aux Green* Frequency doubled auxiliary field

The mode cleaner cavity consists of three mirrors mounted on an aluminium spacer, with one mirror mounted with a piezo-electric transducer (PZT) to actuate the cavity path length. The MCC was locked to resonance using the PDH locking technique (Sect. 5.6.3) with photodetector PDH-MCC as the control loop sensor.

The SHG is a custom built system from Innolight GmbH (Diabolo model [36]), including a PDH cavity length controller. The nonlinear medium is a 5 % magnesium-oxide doped lithium niobate (MgO:LiNbO$_3$) hemilithic crystal, combined with a PZT mounted mirror to form an optical cavity.

The pump field, after propagating through the Mach-Zehnder interferometer (MZ), is incident on the Optical Parametric Oscillator (Sect. 6.2) to drive the nonlinear optical process to generate the squeezed vacuum states. The pump field is also used control the OPO cavity length via the PDH locking technique, using photodetector PDH-OPO. Once at resonance, the OPO transmission photodetector (OPO Trans) signal is then used as a DC Voltage Subtraction (Sect. 5.6.1) locking signal for controlling the Mach-Zehnder interferometer. This enabled further stabilisation of the pump power incident on the OPO, needed for the medium-term squeezing stability measurements.

The Local Oscillator power incident on the homodyne beamsplitter was ~1.9 mW, resulting in the dark noise clearance of greater than 20 dB. The generated vacuum squeezed state enters the homodyne detector signal port.

The modified coherent locking scheme (Sect. 6.3.2) was implemented. Light from the Auxiliary laser (AUX—Mephisto Innolight GmbH), was single-pass frequency doubled with a standalone SHG (SHG2), consisting of a PPKTP crystal in a temperature-controlled oven (no optical cavity). After attenuation of the auxiliary field optical power via dichroic mirrors, these co-propagating auxiliary fields were incident on the dichroic beamsplitter (DBS—HR 1064 nm/AR 532 nm) in the pump field path. The beat note of the two green fields at 59.6 MHz was detected at the 'F-lock' photodetector. This signal was fed back to the Auxiliary laser using Offset frequency locking (Sect. 5.6.2), maintaining a Main-Auxiliary laser frequency offset of 29.8 MHz. The phase relationship between the reflected auxiliary field and pump field was now fixed at the dichroic beamsplitter. The attenuated auxiliary field entered the OPO cavity via the input coupler, sampling the nonlinear interaction. After exiting the OPO, the field co-propagated with the squeezed field to the homodyne detector, where the second coherent sideband control loop was measured.

## 6.8 Results from the DB-OPO Squeezed Light Source

This section reports the results from the DB-OPO squeezed light source. This section begins with a summary of the parameters for the experiments. The three experiments measuring squeezing ellipse phase noise, squeezing time series data and squeezing spectrum data are presented. The results are characterised in terms of the properties affecting the squeezing magnitude and low frequency measurement, and placed into the apparatus upgrade progression. An inference of the squeezing magnitude leaving the OPO concludes this section.

### *6.8.1 Summary of Measurement Parameters*

The common-mode rejection ratio and measurement efficiencies for the three different experiments are listed in Table 6.3.

A common-mode rejection ratio of up to 80 dB was consistently achieved for the homodyne LO port. With the CMR optimised for the local oscillator port, a CMR of

**Table 6.3** Common-mode rejection ratio and efficiency values for the different experiments (Squeezing time series, squeezing ellipse phase noise and squeezing spectrum measurements)

| Parameter | Measurements | | |
|---|---|---|---|
| | Cont. time | Phase noise | Spectrum |
| CMR (LO) (dB) | $75 \pm 5$ | $75 \pm 5$ | $75 \pm 5$ |
| CMR (signal) (dB) | $45 \pm 5$ | $45 \pm 5$ | $45 \pm 5$ |
| $\eta_{prop}$ | $99.3 \pm 0.2$ | $\mathbf{97.3 \pm 0.2}$ | $99.3 \pm 0.2$ |
| $\eta_h$ | $\mathbf{96.5 \pm 0.3}$ | $98.2 \pm 0.3$ | $98.8 \pm 0.3$ |
| $\eta_{pd}$ | $\mathbf{94.0 \pm 1.0}$ | $98.0 \pm 1.0$ | $98.0 \pm 1.0$ |
| $\eta_{det}$ | $90.1 \pm 1.0$ | $93.6 \pm 1.0$ | $96.2 \pm 1.0$ |
| $\eta_{esc}$ | $98.5 \pm 0.3$ | $98.5 \pm 0.3$ | $98.5 \pm 0.3$ |
| $\eta_{tot}$ | $88.7 \pm 1.0$ | $92.2 \pm 1.0$ | $94.7 \pm 1.0$ |

The efficiency values are listed in terms of percentages. Highlighted are the notable efficiency differences between the measurements. The squeezing spectrum measurement had the best detection efficiency of the three measurements

up to 50 dB was obtained for the homodyne signal port. This lower CMR for the signal port is not ideal, however, with lower optical power in the coherent sideband locking field relative to the Local Oscillator field power, a CMR of 50 dB was adequate.

The variations in the measurement efficiencies are highlighted in Table 6.3. More explicitly:

- For the time series measurement, the squeezer was in the April 2011 configuration where the new high-efficiency photodiodes were not yet installed. The loss of squeezing magnitude from the then maximum squeezing of 8.6 dB is attributed to a loss in the homodyne detection efficiency $\eta_h$.
- With the squeezing ellipse phase noise measuremment, the configuration of the squeezer was the November 2011 configuration, with a temporary change in the squeezing signal path optics. This introduced a temporary 2 % reduction in the propagation efficiency.
- The measurement efficiency was maximised for the squeezing spectrum measurement. This was the November 2011 configuration with no extra optical losses.

### 6.8.2 *Squeezing Phase Noise Measurement*

A squeezing/antisqueezing measurement as a function of OPO nonlinear gain (Sect. 6.4.5) was undertaken to assess the total measurement efficiency and the squeezing ellipse phase noise. The result is shown in Fig. 6.15.

The total measurement efficiency $\eta_{tot}$ fitted from the anti-squeezing data was $92.8 \pm 0.4$ %. This agrees well with the experiment parameters shown in Table 6.3. From the squeezing data, the squeezing-ellipse rms-phase noise is fitted to be $11 \pm 2$ mrad.

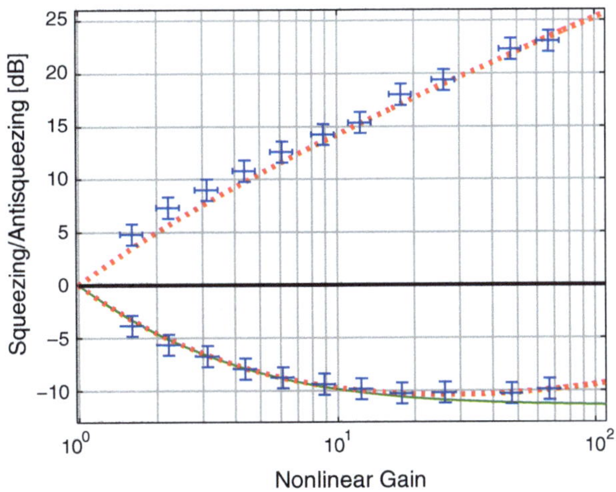


**Fig. 6.15** Squeezing and antisqueezing magnitude, as a function OPO nonlinear parametric gain. *Dotted curves* represent the fit to the data giving an rms-phase noise of $\tilde{\theta} = 11 \pm 2$ mrad. The *solid trace* in the squeezing regime is a theoretical curve with no phase fluctuations ($\tilde{\theta} = 0$ mrad)

### *6.8.3 Time Series Measurement*

The time-series measurement was undertaken to measure the medium-term stability of the produced squeezing magnitude from the OPO. For the time series measurement, the homodyne detector output was connected to a National Instruments PCI-4464 Dynamic Signal Acquisition Card, operated with the National Instrument LabVIEW program. The Mach-Zehnder interferometer on the squeezer was active for this measurement. This was needed to stabilise the pump power to the OPO, minimising photothermal fluctuations in the crystal temperature. The data signal was sampled at 3 kHz, and collected over 5900 s. Converting the data into the Fourier domain in data post-processing, the squeezed time-series shown in Fig. 6.16 were obtained.

From the 500 Hz measurement data, 5900 s of continuous data of squeezing at $7.5 \pm 0.5$ dB was obtained. The low frequency 4 Hz data band has prominent spike-glitches, while the 500 Hz data is clean. This is due to non-stationary noise affecting low frequency measurement (Sect. 6.5.2), as the traversing dust particulates cause short-lived but large reductions in the CMR, resulting in added intensity noise. The impact of non-stationary noise is shown in Fig. 6.17, alongside a spectrum from a period within the data run with no impact of non-stationary noise. This clean spectrum presents flat squeezing of 7.5 dB down to 1 Hz measurement frequency.

Squeezed-light sources with demonstrated operation over 20 h, and duty cycles over 97 %, have been achieved. These sources use dedicated data acquisition, and specialised electronic hardware for auto-relocking sequences [15]. Longer measurement runs with the squeezer were beyond the available hardware at the time, however, with upgrades in control electronics to similar as those in [15], the DB-OPO squeezed light source should be operable on similar timescales.

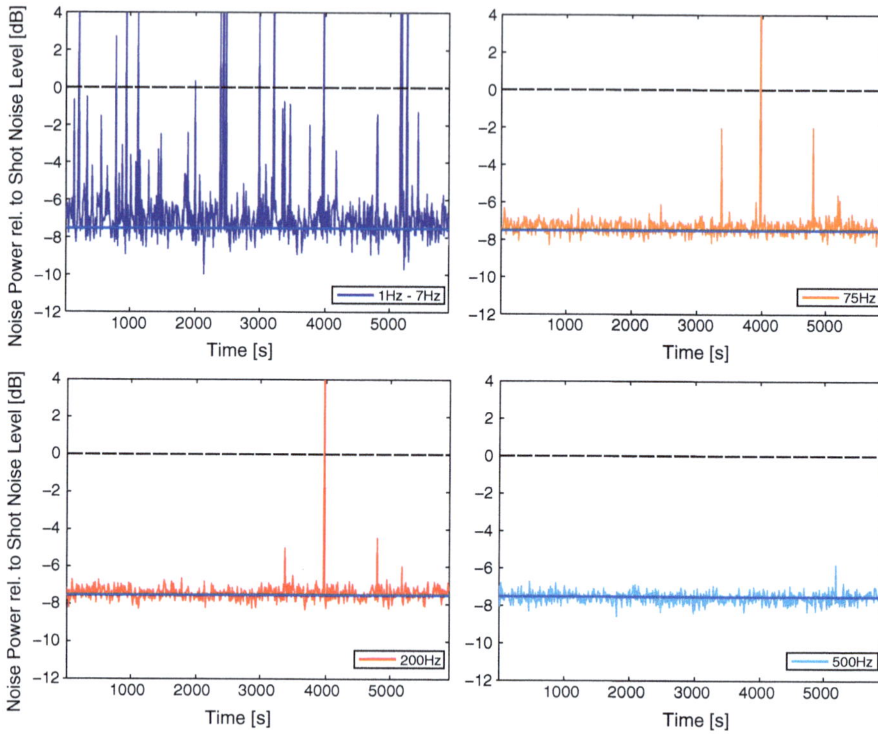


**Fig. 6.16** Squeezing magnitude measurements over 5900 s, normalised to the shot noise level. The measurement bins are $4 \pm 3$, $75 \pm 20$, $200 \pm 20$ and $500 \pm 20$ Hz. The squeezing level is $7.5 \pm 0.5$ dB, commensurate with experiment parameters of that measurement. The *spike-glitches* indicate traversing dust particulates in the homodyne arms causing non-stationary noise at low frequencies

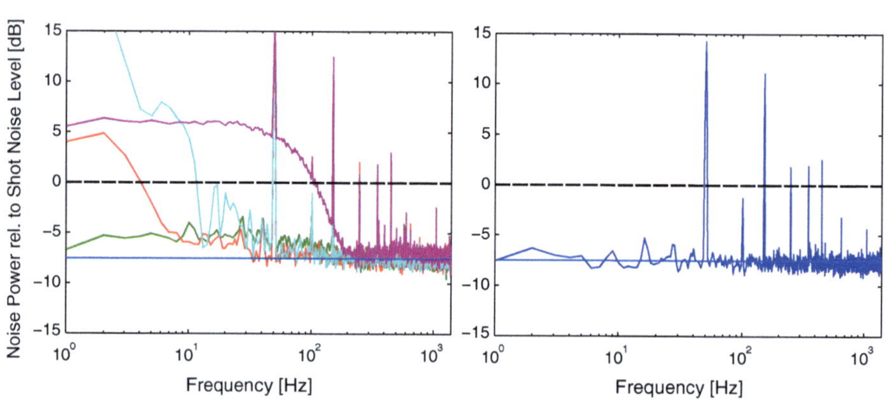


**Fig. 6.17** Post-processed spectra from the time-series measurement. All traces are averaged over 100 s of data. *Left* Non-stationary noise polluting squeezing measurement at low frequencies. *Right* Flat squeezing of 7.5 dB down to 1 Hz. The peaks are due to coupling of AC 50 Hz electric mains harmonics

### 6.8.4 First Measurement of Greater than 10 dB Squeezing Across the Audio Gravitational-Wave Detection Band

Figure 6.18 shows measurements of the electronic dark noise, shot noise, squeezing and antisqueezing spectra between 10 Hz and 100 kHz, made with the DB-OPO squeezer in November 2011 configuration. These measurements were taken by a Stanford Research Systems SRS785 Spectrum Analyzer. All traces were taken with 100 rms averages.

The peaks in the electronic dark noise are attributed to AC mains harmonics (50, 150 and 250 Hz) and electronic ground loop coupling (20 kHz and above). The electronic dark noise has not been subtracted from the shot noise, squeezing or antisqueezing traces. The shot noise level is spectrally flat, and broadly clears the electronic dark noise level by greater than 20 dB.

From the squeezing spectrum, $11.6 \pm 0.4$ dB of squeezing is measured down to 200 Hz, and greater than 10 dB of squeezing is observed between 10 and 200 Hz. This is the first measurement of greater than 10 dB squeezing across the audio gravitational-wave detection band. This measurement demonstrates the capability of generating large squeezing magnitudes applicable for gravitational-wave detector enhancement. The anti-squeezing level is $17.5 \pm 0.4$ dB above the shot noise and is spectrally flat.

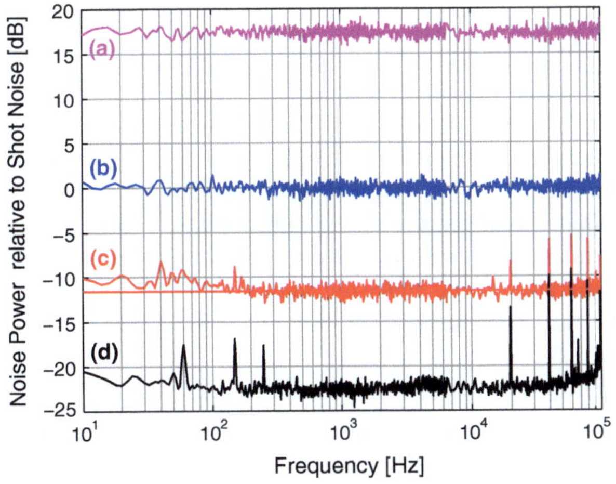


**Fig. 6.18** First measurement of greater than 10 dB squeezing across the audio gravitational-wave detection band, with $11.6 \pm 0.4$ dB down to 200 Hz. Traces are **a** anti-squeezing, **b** shot noise, **c** squeezing and **d** dark noise. A *horizontal line* shows the average squeezing level between 1.6 to 6.4 kHz of 11.6 dB, consistent with the level 200 Hz and above. All traces are formed from 3 FFT windows: 0–1.6, 0–6.4, 0–102.4 kHz with 800 FFT lines (resolution bandwidths of 2, 8, and 128 Hz respectively). All traces taken with 100 rms averages. Dark noise has not been subtracted from the other traces. Reprinted with permission from [25], copyright (2012) by IOP Publishing

The squeezing magnitude is reduced from 200 Hz and below. This is due to the presence of backscattered light interference that could not be mitigated with beam blocks from the measurement. This shows that backscattered noise effects were suppressed to a level approximately 10 dB below the shot noise level.

### 6.8.5 *Inferred Squeezing Magnitude*

Using the measured squeezing and anti-squeezing levels of the spectrum measurement, the squeezing level exiting the OPO can be inferred. This inferred value provides an estimate to the amount of squeezing injected into a gravitational-wave detector, as the interferometer becomes the squeezed state detector.
The inference process corrects for:

- The dark noise contribution, a small contribution given the dark noise clearance.
- The squeezing ellipse phase noise. This uses Eq. 6.17, and the measured values of squeezing phase noise ($11 \pm 2$ mrad) and antisqueezing level ($17.5 \pm 0.4$ dB).
- The detection efficiency, $\eta_{det} = 96.2 \pm 1.0\,\%$, using Eq. 6.6.

The squeezing level corrected for dark noise and squeezing ellipse phase noise is $12.4 \pm 1.0$ dB. Correcting for detection efficiency, the inferred squeezing magnitude level exiting the OPO is $17.0 \pm 0.3$ dB.

## 6.9 Summary

In this chapter, the doubly-resonant, travelling-wave Optical Parametric Oscillator, and its design reasoning were presented, followed by the modified coherent sideband locking scheme used for squeezing ellipse phase control. The properties that affect squeezing magnitude and low-frequency measurement were then detailed. The progression of upgrades to the squeezed light source, reflecting the progression of knowledge gained into these properties, was outlined. The schematic of the DB-OPO squeezer was then shown.
Results of squeezing measurements were presented:

- The time-series measurement showed continuously-controlled squeezing of $7.5 \pm 0.5$ dB over 5900 s. This measurement highlighted the impact of non-stationary noise in low-frequency measurement, and also provided a post-processed spectrum of flat squeezing of 7.5 dB down to 1 Hz measurement frequency.
- A measurement of the squeezing ellipse phase noise from the squeezed light source resulted in $11 \pm 2$ mrad rms-phase noise.
- the first measurement of greater than 10 dB squeezing across the audio gravitational-wave detection band, $11.6 \pm 0.4$ dB above 200 Hz. The loss of squeezing magnitude below 200 Hz in the spectrum measurement was due to the backscattered light interferences that could not be suppressed.

- Combining experiment efficiency parameters with the squeezing ellipse phase noise measurement resulted in an inferred squeezing magnitude level exiting the DB-OPO of $17.0 \pm 0.3$ dB.

## References


1. Y. Takeno, M. Yukawa, H. Yonezawa, A. Furusawa, Observation of −9 dB quadrature squeezing with improvement of phase stability in homodyne measurement. Opt. Express **15**, 7 (2007)
2. T. Eberle, S. Steinlechner, J. Bauchrowitz, V. Händchen, H. Vahlbruch, M. Mehmet, H. Müller-Ebhardt, R. Schnabel, Quantum enhancement of the zero-area sagnac interferometer topology for gravitational wave detection. Phys. Rev. Lett. **104**, 251102 (2010)
3. H. Yonezawa, K. Nagashima, A. Furusawa, Generation of squeezed light with a monolithic optical parametric oscillator: Simultaneous achievement of phase matching and cavity resonance by temperature control. Opt. Express **18**(19) (2010)
4. B.E.A. Saleh, M.C. Teich, *Fundamentals of Photonics* (Wiley, New York, 1991)
5. G.D. Boyd, D.A. Kleinman, Parametric interaction of focussed gaussian light beams. J. Appl. Phys. **39**(8) (1968)
6. D.A.B. Miller, A.C. Gossard, W. Wiegmann, Optical bistability due to increasing absorption. Opt. Lett. **9**, 162 (1984)
7. R. Boyd, *Nonlinear Optics* (Academic Press, San Diego, 1992)
8. S. Pearl, H. Lotem, Y. Shimony, S. Rosenwaks, Optimization of laser intracavity second-harmonic generation by a linear dispersion element. J. Opt. Soc. Am. B **16**(10), 1705–1711 (1999)
9. K. McKenzie, M. Gray, S. Goßler, P.K. Lam, D. McClelland, Squeezed state generation for interferometric gravitational-wave detection. Class. Quantum Gravity **23**, S245–S250 (2006)
10. G. Imeshev, M. Proctor, M.M. Fejer, Phase correction in double-pass quasi-phase-matched second-harmonic generation with a wedged crystal. Opt. Lett. **23**(3), 165–167 (1998)
11. I. Juwiler, A. Arie, A. Skliar, G. Rosenman, Efficient quasi-phase-matched frequency doubling with phase compensation by a wedged crystal in a standing-wave external cavity. Opt. Lett. **24**(17), 1236–1238 (1999)
12. N. Grosse, W.P. Bowen, K. McKenzie, P.K. Lam, Harmonic entanglement with second-order nonlinearity. Phys. Rev. Lett. **96**, 063601 (2006)
13. M. Stefszky, C.M. Mow-Lowry, K. McKenzie, S. Chua, B.C. Buchler, T. Symul, D.E. McClelland, P.K. Lam, An investigation of doubly-resonant optical parametric oscillators and nonlinear crystals for squeezing. J. Phys. B: At. Mol. Opt. Phys. **44**, 015502 (2011)
14. K. McKenzie, E. Mikhailov, K. Goda, P.K. Lam, N. Grosse, M. Gray, N. Mavalvala, D. McClelland, Quantum noise locking. J. Opt. B: Quantum Semiclass. Opt. **7**, S421–S428 (2005)
15. A. Khalaidovski, H. Vahlbruch, N. Lastzka, C. Gräf, K. Danzmann, H. Grote, R. Schnabel, Long-term stable squeezed vacuum state of light for gravitational wave detectors. Class. Quantum Gravity **29**, 075001 (2012)
16. S. Chelkowski, H. Vahlbruch, K. Danzmann, R. Schnabel, Coherent control of broadband vacuum squeezing. Phys. Rev. A **75**, 0043814 (2007)
17. H. Vahlbruch, S. Chelkowski, B. Hage, A. Franzen, K. Danzmann, R. Schnabel, Coherent control of vacuum squeezing in the gravitational-wave detection band. Phys. Rev. Lett. **97**, 011101 (2006)
18. S.S.Y. Chua, M. Stefszky, C. Mow-Lowry, B. Buchler, S. Dwyer, D. Shaddock, P.K. Lam, D. McClelland, Backscatter tolerant squeezed light source for advanced gravitational-wave detectors. Opt. Lett. **36**(23), 4680 (2011)
19. S.S.Y. Chua, B.J.J. Slagmolen, D.A. Shaddock, D.E. McClelland, Quantum squeezed light in gravitationalwave detectors. Class. Quantum Gravity **31**, 183001 (2014)

20. P.K. Lam, T. Ralph, D. McClelland, B. Buchler, H.-A. Bachor, J Gao, Optimization and transfer of vacuum squeezing from an optical parametric oscillator. J. Opt. B: Quantum Semiclass. Opt. **1**, 469–474 (1999)
21. T. Aoki, G. Takahashi, A. Furusawa, Squeezing at 946 nm with periodically poled $KTiOPO_4$. Opt. Express **14**, 6930–6935 (2006)
22. A. Franzen, B. Hage, J. DiGuglielmo, J. Fiurášek, R. Schnabel, Experimental demonstration of continuous variable purification of squeezed states. Phys. Rev. Lett. **97** 150505 (2006)
23. K. McKenzie, N. Grosse, W. Bowen, S. Whitcomb, M. Gray, D. McClelland, P.K. Lam, Squeezing in the audio gravitational-wave detection band. Phys. Rev. Lett. **93**(16), 161105 (2004)
24. H. Vahlbruch, S. Chelkowski, K. Danzmann, R. Schnabel, Quantum engineering of squeezed states for quantum communication and metrology. New J. Phys. **9**, 371 (2007)
25. M. Stefszky, C. Mow-Lowry, S. Chua, D. Shaddock, B. Buchler, H. Vahlbruch, A. Khalaidovski, R. Schnabel, P.K. Lam, D. McClelland, Balanced homodyne detection of optical quantum states at audio-band frequencies and below. Class. Quantum Gravity **29**, 145015 (2012)
26. P. Horowitz, W. Hill, *The Art of Electronics*, 2nd edn. (Cambridge University Press, Cambridge, 1991)
27. F. Seifert, Power stabilization of high power lasers for second generation gravitational wave detectors. Ph.D. thesis, Gottfried Wilhelm Lebniz Universität, Hannover, Germany, 2010
28. P. Kwee, Laser characterization and stabilization for precision interferometry. Ph.D. thesis, Gottfried Wilhelm Leibniz Universität, Hannover, Germany, 2010
29. S. Chua, K. McKenzie, B.C. Buchler, D.E. McClelland, Impact of non-stationary events on low frequency homodyne detection. J. Phys.: Conf. Ser. **122**, 012023 (2008)
30. Raicol Crystals Ltd., http://www.raicol.com/
31. Photon LaserOptik GmbH, http://www.photon-laseroptik.de/
32. LaserOptik GmbH, http://www.laseroptik.de/
33. H. Vahlbruch, Squeezed light for gravitational wave astronomy. Ph.D. thesis, Albert Einstein Institute and Leibniz University of Hannover, Hannover, Germany, 2008
34. ATFilms, http://www.atfilms.com/
35. T. Day, A.C. Nilsson, M.M. Fejer, A.D. Farinas, E.K. Gustafson, C.D. Nabors, R.L. Byer, 30 Hz-linewidth, diode-laser-pumped, Nd:GGG nonplanar ring oscillators by active frequency stabilisation. Electron. Lett. **25**, 810 (1989)
36. Innolight GmbH, http://www.innolight.de/

# Chapter 7
# Backscatter Tolerance of a Travelling-Wave Optical Parametric Oscillator

**Abstract** Backscattered light in the squeezing propagation path is a source of low frequency noise. In a gravitational-wave detector with squeezing, this can potentially be a sensitivity-limiting noise source (The influence of backscattered light with squeezed-interferometric readout will be covered in Chap. 10.). A possible back-reflecting source in the squeezing path is the OPO cavity itself. As stated in Chap. 6, the travelling-wave design of the DB-OPO was chosen because of its first-order immunity to backscattered light. This chapter presents analysis and experiment results that quantifies the tolerance of a travelling-wave OPO squeezing cavity to backscattered light. Reported as part of (Chua et al., Opt Lett 36(23):4608, 2011, [1]), this is to the author's knowledge the first such measurement of its kind. After defining backscattered light, the theoretical framework for calculating the backscatter reflectivity of the OPO is introduced. The experiments performed to determine the intrinsic isolation of a travelling-wave OPO report an isolation factor of $(41 \pm 2)$ dB. This is followed by a calculation of the bi-directional scatter distribution function for the OPO. A discussion on the determined backscatter tolerance value concludes the Chapter.

## 7.1 Backscattered Light

Backscattered light is scattered light that is subsequently reflected back towards the measurement output, optically re-interfering with the measurement Local Oscillator. This light can spuriously add noise to measurements [1–3].

Possible backscatter objects in the squeezing path include optical mount surfaces, scattering centres, and the OPO cavity itself via direct back-reflection. For a standing-wave cavity OPO (Sect. 6.1.1), there is no distinction between forward-propagating and reverse-propagating directions. Thus to mitigate incident scattered light, at least one Faraday isolator is required in the squeezing path (two Faradays in series were used in [4]). Even with current state-of-the-art components, Faraday isolators add optical loss (Sect. 6.4.1) to the squeezed field, typically between 2–10 %, thus reducing the squeezing magnitude available.

 
S.S.Y. Chua, *Quantum Enhancement of a 4 km Laser Interferometer Gravitational-Wave Detector*, Springer Theses,
DOI 10.1007/978-3-319-17686-4_7

By using an OPO with inherent isolation to backscattered light, such as a travelling-wave bow-tie configuration, this would reduce the need for additional Faraday isolators. With a traveling-wave bow-tie OPO, scattered light ideally does not reflect back toward the Local Oscillator. This is from the angular separation of input and reflected/output beams. In practice, there is coupling between the reverse and forward propagating fields inside the OPO, caused by imperfections of the interfaces within the cavity. Experiments to quantitatively determine this backscatter coupling or reflectivity are reported here.

## 7.2 Theoretical Analysis of an OPO Cavity Backscattered-Light Reflectivity

The goal is to determine an expression for the reflectivity of back-propagating power incident on the OPO, namely $R_r = P_F^A / P_B^A$. The reflectivity $R_r$ represents the *intrinsic isolation of the OPO cavity to backscattered light*. This section constructs a theoretical model for determining the intrinsic isolation.

We begin by considering the OPO cavity shown in Fig. 7.1. The back-propagating field $A_B$ is the only incident field on the cavity. This field enters the cavity at the input/output coupler, with coupling rate $\kappa_{in}$. The other cavity mirrors have coupling rates $\kappa_{2,3,4}$ respectively. The coupling between reverse and forward-propagating intra-cavity modes is accounted for with the *intra-cavity coupling rate* $\kappa_c$. This is assumed to completely occur at the nonlinear crystal, the sole transmissive optic in the OPO, where the small cavity beam waist is located. The total loss rate from the cavity is dominated by the cavity mirror losses, thus $\kappa \approx \sum \kappa_{in,2,3,4}$. We utilise the quantum Langevin cavity equations (Eq. 3.51), considering only the steady-state terms. The cavity field equations can be expressed as

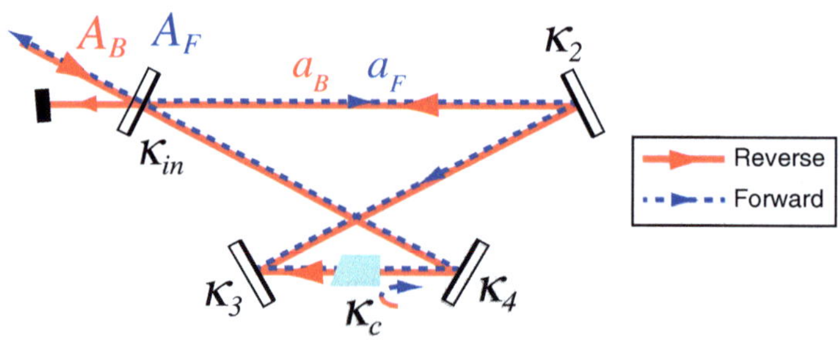


**Fig. 7.1** Diagram of an OPO cavity for a theoretical backscatter-reflectivity calculation. $\kappa_{in}$ input/output coupling rate, $\kappa_{2,3,4}$ cavity mirrors' coupling rates; $\kappa_c$ intra-cavity mode coupling rate; $a_B$, $a_F$ reverse (*red solid*) and forward (*blue dashed*) propagating intra-cavity modes; $A_B$, $A_F$ reverse-and forward-propagating external fields. The forward and reverse field directions are coloured separately

$$\dot{a}_F = -\kappa a_F + \sqrt{2\kappa_c} a_B - \sqrt{2\kappa_c} a_F \tag{7.1}$$

$$\dot{a}_B = -\kappa a_B + \sqrt{2\kappa_{in}} A_B - \sqrt{2\kappa_c} a_B + \sqrt{2\kappa_c} a_F \tag{7.2}$$

These are solved for the steady-state solution ($\dot{a}_F = 0,\ \dot{a}_B = 0$), resulting in forward ($a_F$) and reverse ($a_B$) intra-cavity fields

$$a_F = \frac{\sqrt{4\kappa_c \kappa_{in}}}{(\kappa_c + \sqrt{2\kappa_c})^2 - 2\kappa_c} A_B \tag{7.3}$$

$$a_B = \frac{\sqrt{2\kappa_{in}}(\kappa_c + \sqrt{2\kappa_c})}{(\kappa_c + \sqrt{2\kappa_c})^2 - 2\kappa_c} A_B \tag{7.4}$$

From Eq. 7.3, we express

$$\left|\frac{a_F}{A_B}\right|^2 = \left(\frac{\sqrt{4\kappa_c \kappa_{in}}}{(\kappa_c + \sqrt{2\kappa_c})^2 - 2\kappa_c}\right)^2 \tag{7.5}$$

Now, recalling the definitions for power of external optical fields (Sect. 3.5.1), and the cavity input-ouput relations (Eqs. 3.62 and 3.63),

$$P_F^A = \hbar\omega_A 2\kappa_{in} |a_F|^2, \quad P_B^A = \hbar\omega_A |A_B|^2 \tag{7.6}$$

and substituting in Eq. 7.5, the reflectivity of back-propagating light power incident on the optical cavity to the forward-propagating direction $R_r$, is

$$R_r = \frac{P_F^A}{P_B^A} = 2\kappa_{in} \left|\frac{a_F}{A_B}\right|^2 = 2\kappa_{in} \left(\frac{\sqrt{4\kappa_c \kappa_{in}}}{(\kappa_c + \sqrt{2\kappa_c})^2 - 2\kappa_c}\right)^2 \tag{7.7}$$

We now need the intra-cavity coupling rate $\kappa_c$. From Eqs. 7.3 and 7.4, we can write an expression for the *intra-cavity coupling reflectivity* $R_{\text{coup}}$. This is the ratio between reverse and forward propagating intra-cavity mode optical power, namely

$$R_{\text{coup}} = \left|\frac{a_F}{a_B}\right|^2 = \frac{2\kappa_c}{(\kappa + \sqrt{2\kappa_c})^2} \tag{7.8}$$

From the intra-cavity coupling reflectivity $R_{\text{coup}}$, the intra-cavity coupling rate is

$$\kappa_c = \frac{R_{\text{coup}} \kappa^2}{2(1 - \sqrt{R_{\text{coup}}})^2} \tag{7.9}$$

Substituting for $\kappa_c$ into Eq. 7.7 results in

$$R_r = \frac{4\kappa_{in}^2 R_{\text{coup}}}{\kappa^2 (1 + 3R_{\text{coup}})} \tag{7.10}$$

Therefore, by knowing the input coupler rate $\kappa_{in}$, the total cavity loss rate $\kappa$, and the intra-cavity coupling reflectivity $R_{\text{coup}}$, the power reflectivity of the OPO to backscattered light can be determined.

The derivation has been based on a four-mirror cavity in bow-tie configuration. However, Eq. 7.10 can be applied to standing-wave cavities. This is achieved by adjusting the assignment of cavity loss rates to equivalent mirrors, and noting the intra-cavity coupling reflectivity is unitary ($R_{\text{coup}} = 1$).

## 7.3 Determining the Intrinsic Isolation $R_r$

Direct measurement of the intrinsic isolation $R_r$ is challenging on two fronts. Firstly as the reflected beam of interest is overlapped with the input 'scattered' beam, and secondly the reflected field is of low optical power. To work around the input/reflected beam overlap, we focus on a measurement of the intra-cavity coupling reflectivity $R_{\text{coup}}$. Thus for the bow-tie geometry, we can inject the input reverse beam into another cavity port.

With the reflected field's low optical power, the balanced homodyne detector is a component ideal for making such a measurement. We now consider the derivation of the homodyne measurement of $R_{\text{coup}}$.

### *7.3.1 Balanced Homodyne Measurement to Determine $R_{\text{Coup}}$*

Consider a homodyne measurement on an optical cavity transmission signal $\mathcal{T}(\omega)$ (defined in Sect. 3.5.2). The homodyne detector will measure the amplitude or phase quadrature of the transmission ($|\mathcal{T}(\omega)|^2$ and $\angle|\mathcal{T}(\omega)|^2$ respectively), depending on the relative phase $\Delta\phi$ of the transmission signal to the local oscillator field. If we scan the cavity transmission profile (via changing the cavity optical path length) whilst sweeping the relative phase, the homodyne will measure a convolution between amplitude and phase quadrature responses of the cavity transmission signal. The result of this *dual-scanning measurement* is shown in Fig. 7.2a.

From the difference photocurrent of the balanced homodyne (Eq. 5.38), we now write

$$V_{\text{trans}} = i_{-}R = 2A_F B\cos(\Delta\phi)R \tag{7.11}$$

$$P_F^A = \hbar\omega_A A_F^2 = \hbar\omega_A 2\kappa_{in}|a_F|^2 = \hbar\omega_A\left(\frac{V_{\text{trans}}}{2BR}\right)^2 \tag{7.12}$$

where $V_{\text{trans}}$ is the signal voltage measured on the homodyne, $A_F$ is the transmission signal field amplitude, $B$ is the LO field amplitude and $P_F^A$ is the cavity transmission signal power. We note for the various parameters:

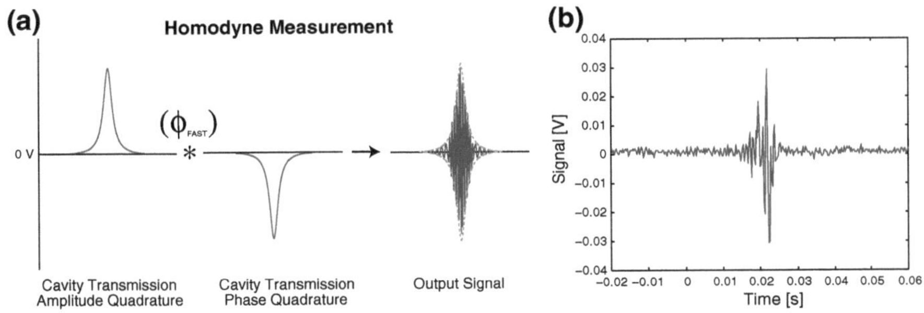


**Fig. 7.2** Dual-scanning measurement with the homodyne to determine $R_{\text{coup}}$. **a** Illustration of the homodyne *output* for the dual-scanning measurement. The results is a convolution between amplitude and phase quadratures of the cavity transmission response. **b** Measured data of one such cavity resonance sweep

- $B$, the Local Oscillator field component, is given by $B = \sqrt{P_{LO}}$
- $\overline{\cos^2 \Delta\phi} = 1/2$, which results from the signal spending equal time in each measurement quadrature as the relative phase is swept over many cycles.
- $R$ is the electronic transimpedance of the homodyne detector. With measurement of optical power $P_{spd}$ and voltage $V_{spd}$ on a single homodyne photodetector, we can determine $R = V_{spd}^2/(\sqrt{2} \times P_{spd})$, using Ohm's Law and correcting for the presence of two single detectors in the homodyne [5].

Thus applying the above parameters, and correcting for the cavity input/output coupler rate, a measurement of the forward intra-cavity mode power $P_F^a$ can be made using the homodyne detector.

### 7.3.2 Experimental Layout and Result

The experiment setup for measuring the intra-cavity coupling reflectivity $R_{\text{coup}}$ is shown in Fig. 7.3. A reverse beam (RV) was coupled into the OPO. The pump and auxiliary field beams were blocked, thus the OPO was isolated from all other input beams. The cavity length was modulated with a ramp signal of 2 Hz (SLOW), whilst the homodyne detector Local Oscillator phase was modulated with a ramp signal at 2 kHz (FAST). The reverse beam power was measured with an optical power meter (Newport 841-PE) before each injection. Many sweeps of the cavity transmission signal were measured on an oscilloscope (Agilent 54624A) for a given input reverse-field power. This was then repeated with various input reverse-field powers, adjusted with a variable power attenuator (VA) placed in the reverse beam path. After correcting for the injection mirror coupling rates, the data and resulting trend is shown in Fig. 7.4.

The slope of the data trend is the reverse-to-forward intra-cavity coupling reflectivity. The value determined is $R_{\text{coup}} = (2.3 \pm 0.3) \times 10^{-5}$.

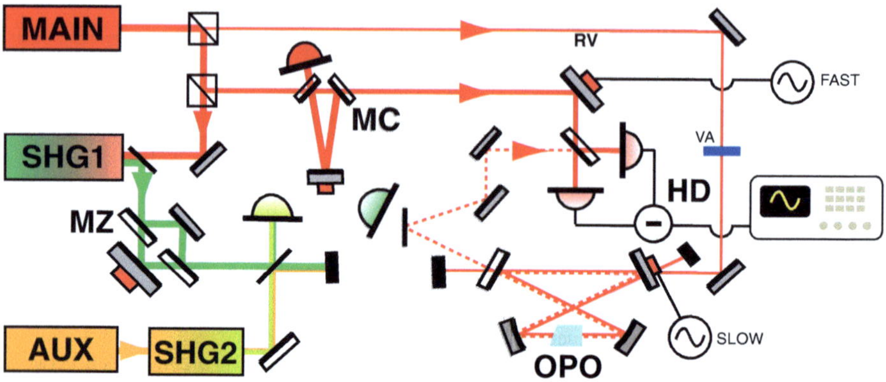


**Fig. 7.3** Experiment setup for the intra-cavity coupling reflectivity ($R_{\mathrm{coup}}$) measurement. *RV* reverse-seed beam (the only beam incident on the OPO); *VA* Variable power attenuator; *SLOW* Slow modulation sweep of the OPO optical path length; *FAST* Fast modulation of the LO path length to sweep the relative phase ($\Delta\phi$) over many cycles

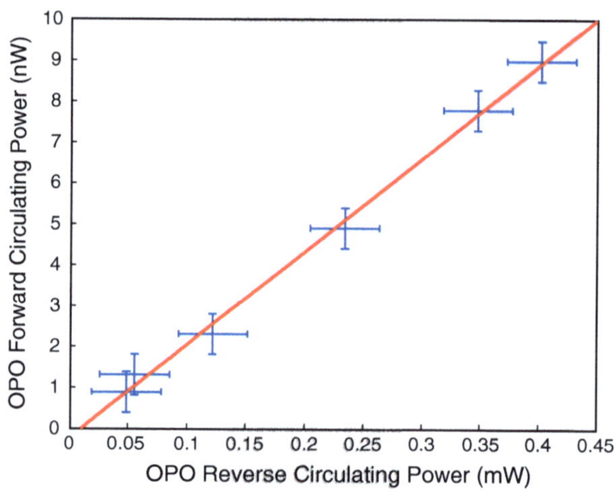


**Fig. 7.4** Measured optical forward intra-cavity power as a function of reverse intra-cavity power. The slope value is $R_{\mathrm{coup}}$, fitted to be $(2.3 \pm 0.3) \times 10^{-5}$. Reprinted with permission from [1], copyright (2011) by the Optical Society of America

Using Eq. 7.10 and the OPO cavity parameters listed in Table 6.1, the intrinsic isolation of the OPO to backscatter is determined to be $R_r = 41 \pm 2\,\mathrm{dB}$. This isolation value can be compared with:

- A standing-wave cavity with equivalent parameters for an input/output coupler and cavity mirror reflectivities. The reverse-to-forward intra-cavity coupling reflectivity is $R_{\mathrm{coup}} = 1$. The intrinsic isolation to backscatter of an equivalent standing wave cavity is calculated to be 0.03 dB. Therefore, this travelling-wave bow-tie cavity configuration has approximately 40 dB greater intrinsic isolation than an equivalent standing-wave geometry.

- Off-the-shelf Faraday isolators with a nominal isolation of ≥30 dB [6]. The travelling-wave bow-tie design has isolation to backscattered light that is at least comparable or better in magnitude.

An intrinsic isolation of $R_r = 41 \pm 2$ dB quantifies the first-order immunity of a travelling-wave bow-tie OPO to backscattered light. This isolation is a property that gives the benefit of a Faraday isolator without the introduction of additional optical loss.

## 7.4 Bi-directional Scatter Distribution Function (BSDF) of the OPO

The *bidirectional scatter distribution function* (*BSDF*) is the measure of back-scattered-light reflectivity for a transmissive optic [7]. This section calculates a value for the BSDF for the travelling-wave OPO. This is appropriate as from the theoretical treatment (Sect. 7.2), the scatter reflectivity of the OPO is centralised at the nonlinear crystal of the OPO. The BSDF is given by [7]

$$\text{BSDF} = \frac{P_{sc}/\Omega_{1/e}}{P_{inc}\cos\theta_s} \tag{7.13}$$

where $\theta_s$ is the input beam's angle of incidence onto the optic, and

$$\Omega_{1/e} = \frac{\lambda^2}{\pi W_0^2} \tag{7.14}$$

is the *solid angle* of the beam [8] at the optic. The small cavity beam waist ($W_0 = 34\ \mu$m) is located at the nonlinear crystal, and both forward- and reverse-propagating fields are at normal incidence, i.e. $\theta_s = 0,\ \cos\theta_s = 1$. Lastly, the OPO cavity finesse needs to be taken into account, thus the power reflectivity of the OPO to backscattered light is related to the BSDF by [9].

$$R_r = \text{BSDF} \times \Omega_{1/e} \times \left(\frac{\mathcal{F}}{\pi\sqrt{R_{in}}}\right)^2 \tag{7.15}$$

Using the measured $R_r$ value and OPO cavity parameters in Table 6.1, the BSDF for the OPO cavity is calculated to be $(1.7 \pm 0.3) \times 10^{-3}$ str$^{-1}$.

The inferred BSDF value can be compared to reflective optical components. The function for describing the backscatter-reflectivity of reflective optics is the *bi-directional reflectance distribution function* (*BRDF*), and the BRDF and BSDF can be approximated as [7]

$$\text{BRDF} \cong \text{BSDF} \tag{7.16}$$

**Table 7.1** Measured BRDF values for optical components

| Company | Optic type | BRDF ($str^{-1}$) |
|---|---|---|
| CVI | Standard high-R mirror for 0°–45° | $5 \times 10^{-4}$ |
| Newport | Nominal R > 0.99, 0°–45° | $1 \times 10^{-4}$ |
| Newport | 50 % non-polarising beamsplitter | $1 \times 10^{-4}$ |

Reproduced from [10]

Measurements of the BRDF for off-the-shelf optical components are shown in Table 7.1. The BSDF of the OPO is approximately ten times larger than the BRDF values of these tested optics. This is a reasonable value when factoring in the multiple optical components that consist of the DB-OPO cavity, along with the two surfaces and bulk material of the nonlinear crystal itself.

## 7.5 Verification of the Intrinsic Isolation with Squeezing Measurement

From the works of McKenzie et al. [11, 12] and Goda et al. [13], a forward seed beam forms a mechanism for low frequency classical noise to couple into the squeezed field. Removing the forward seed beam was the breakthrough in measuring squeezed states in the audio-detection band [11]. This propelled the need for generating squeezed vacuum states, and control techniques with no seed beam present, such as coherent sideband locking (Sect. 6.3.1).

Light backscattered into the forward propagating direction within the OPO can act as a seed beam at the fundamental frequency. By performing low-frequency squeezing measurements with an injected reverse beam, classical noise will enter the measurement as a function of the reverse beam power being reflected into the forward seed-field direction. This provides a verification of the intra-cavity coupling reflectivity $R_{\mathrm{coup}}$.

### *7.5.1 Noise Coupling to Squeezing Measurement*

The theoretical framework used here follows on from the derivation work of McKenzie (Chap. 5 of [12]). The output noise variances of the OPO with classical noise coupling is given by

$$V_{1,2}^{a-out-n} = \underbrace{V_{1,2}^{a-out}}_{(i)} + \frac{2\kappa_{out}^{a}\bar{a}_F^2}{\omega^2 + (\kappa_a \mp \varepsilon\bar{b})^2}\left(\underbrace{\varepsilon^2 V_{1,2}^{b}}_{(ii)} + \underbrace{4\kappa_{in}^{a}\kappa_c V_s^{a}}_{(iii)} + \underbrace{4\bar{b}^2(\mathbf{Re}[\delta\varepsilon])^2}_{(iv)} + \underbrace{4(\mathbf{Re}[\delta\Delta])^2}_{(v)}\right) \tag{7.17}$$

where

$$V_{1,2}^b = \frac{2\kappa_{in}^b V_{1,2}^{B-in} + 2\kappa_{out}^b V_{1,2}^{B-out} + 2\kappa_l^b V_{1,2}^{B-l} + 4\bar{b}^2(\mathbf{Re}[\delta\Delta])^2}{\omega^2 + \kappa_b^2} \tag{7.18}$$

The various components of Eq. 7.17 are

(i) The quantum quadrature-variances, as per Eqs. 5.33 and 5.34
(ii) The coupling of pump field noise $V_{1,2}^b$
(iii) The coupling of reverse-seed field noise $V_s^a$
(iv) The coupling of changes in the phase-matching temperature, causing variations in the nonlinear parameter $\delta\varepsilon$
(v) The coupling of cavity length fluctuations, causing variations in the cavity detuning $\delta\Delta$

The total noise variance consists of a quantum-noise component [(i)], and added classical noise components [(ii)–(v)] scaled by the strength of the forward intra-cavity field $\bar{a}_F^2$.

In the verification experiment reported here, tests of the stability of the crystal temperature controller, reverse-beam intensity noise, pump field noise, and cavity detuning were independently made. From these tests, the dominant classical noise term is the pump field noise term (ii), that is

$$\varepsilon^2 V_{1,2}^b \gg \left[4\kappa_{in}^a \kappa_c V_s^a,\ 4\bar{b}^2(\mathbf{Re}[\delta\varepsilon])^2,\ 4(\mathbf{Re}[\delta\Delta])^2\right] \tag{7.19}$$

Further, the forward-propagating field is the coupling of the reverse-beam into the forward-propagating direction, thus using Eq. 7.8

$$\bar{a}_F^2 = R_{\text{coup}} \times \bar{a}_B^2 \tag{7.20}$$

Therefore, simplifying Eq. 7.17 using Eqs. 7.19 and 7.20, the output noise variances of the OPO become

$$V_{1,2}^{a-out-n} \approx V_{1,2}^{a-out} + \frac{2\kappa_{out}^a \varepsilon^2 R_{\text{coup}} \bar{a}_B^2 V_{1,2}^b}{(\kappa_a \mp \varepsilon\bar{b})^2} \tag{7.21}$$

With measurements of the pump field noise, reverse-beam power and squeezing measurement, the intra-cavity coupling reflectivity $R_{\text{coup}}$ can be verified.

### 7.5.2 Squeezing with Injected Reverse-Beam Field

The schematic of the squeezing measurement with an injected reverse-beam is shown in Fig. 7.5. This is similar to the setup for measurement of the intra-cavity coupling reflectivity $R_{\text{coup}}$, except with the OPO now fully operating. The OPO cavity was

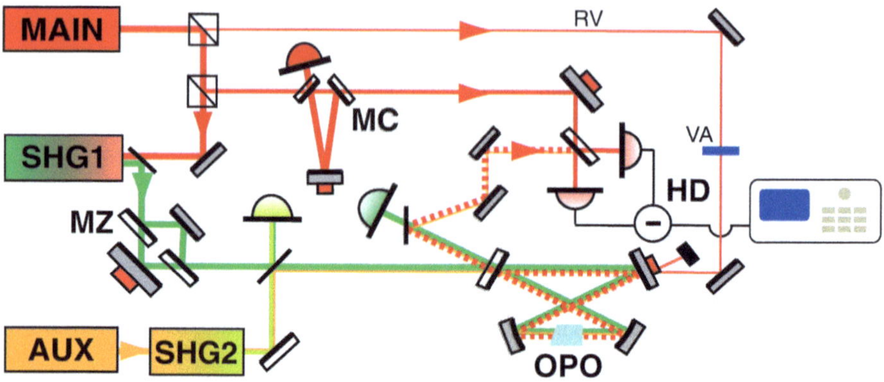


**Fig. 7.5** Schematic of the squeezer with injected reverse-beam. Similar to the setup of measurement for $R_{coup}$ (Fig. 7.3), except with fully operating OPO cavity and no SLOW/FAST modulation

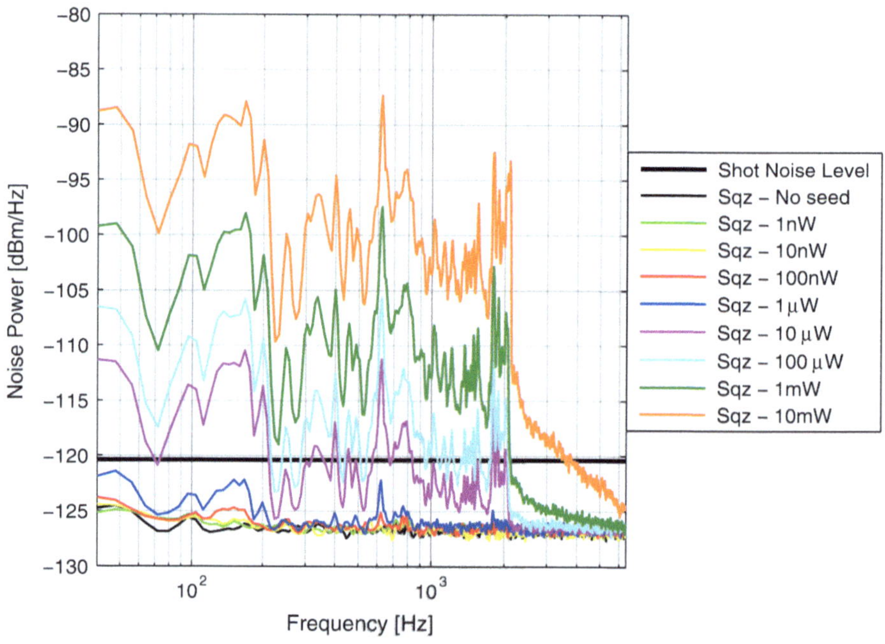


**Fig. 7.6** Squeezing spectra with injected reverse beam for various optical powers. All traces are formed from 1 FFT window 0–6.4 kHz (resolution bandwidth of 8 Hz) with 300 rms averages. Dark noise (22 dB below the shot noise level) has not been subtracted from the traces

operated with 60 mW of input pump power, and the full-modified coherent sideband locking technique was active. The propagation efficiency was 99.3 % and the balanced homodyne detector had a homodyne efficiency of 93.1 ± 0.3 %. Measured spectra are shown in Fig. 7.6, along with the shot noise level. Eight different incident

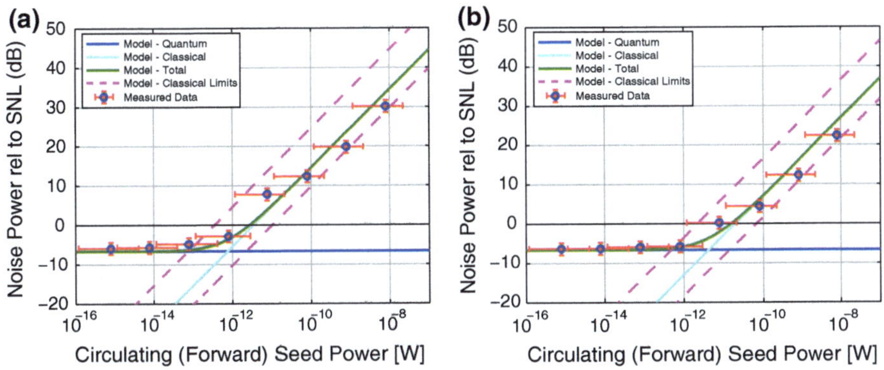


**Fig. 7.7** Integrated noise power of the squeezing spectra measurement, normalised relative to the shot noise level. Curves represent the theoretical model, with the Total curve being the sum of the Quantum and Classical components. Panel **a** 120–180 Hz frequency band and panel **b** 300–400 Hz frequency band. The 'classical limit' curves are discussed in Sect. 7.6

reverse-beam powers between 1 nW and 10 mW were injected. No experimental parameters other than the input seed power were varied during this measurement run.

At low reverse beam powers, the squeezing spectrum is 6 dB flat down to 80 Hz. The spectrum shows features appearing below 150 Hz and around 600 Hz from 100 nW reverse-beam power injected. As the reverse-beam power was increased further, the noise floor and features continued to increase.

The noise power increase is evident in Fig. 7.7. This shows the integrated noise power of the data in Fig. 7.6, in narrow frequency bands. The data is normalised to the shot noise level, and shown as a function of reverse-beam power. Panel (a) is the frequency band between 120 and 180 Hz, and panel (b) is the frequency band between 300 and 400 Hz. The 'Total' model line is the prediction of Eq. 7.21, using the OPO cavity parameters and independent measurements of the input pump noise and cavity detuning (needed for Eq. 7.18). The data follows the upward trend of the theoretical model with the determined value of $R_{\text{coup}}$.

## 7.6 Discussion

### *7.6.1 Uncertainty in $R_{\text{Coup}}$ and Improving Measurements*

There are precision limitations in the measurements performed to determine and verify the intrinsic isolation of the traveling-wave OPO.

- For the dual-scanning measurement used to determine $R_{\text{coup}}$, there is firstly digitisation error from the measurement oscilloscope. The peaks of the forward signal may be undersampled, resulting in an overestimate of $R_{\text{coup}}$. An average over repeated measurements of the cavity transmission signal addressed this limitation.

Secondly there is uncertainty with the input reverse beam power. The variable attenuator, a combination of polarisation optics (Glan-Taylor prism, half-wave and quarter-wave plates) and optical neutral-density filters, is a non-trivial potential source for misalignment of the input beam into the cavity. Beam-alignment stability was only able to be verified at higher input powers.

- The squeezing-reverse-beam measurement has uncertainty in the effective forward-propagating beam power, due to the OPO nonlinear parametric gain. With the relative phase of the forward field and pump field not being controlled, this means that the intra-cavity power of the forward field will time-vary, depending on the amplification or deamplification condition. Recalling from Sect. 5.3.5, Eq. 5.17, the output power accounting for OPO nonlinear gain is given by

$$\frac{P_{\text{out}}^{\pm}}{P_{\text{out}}^{0}} = \left(\frac{1 \pm x}{1 - x^2}\right)^2 \tag{7.22}$$

where $x$ is the normalised pump parameter. The normalised pump parameter is related to the OPO classical nonlinear gain $G$ by [14]

$$x = 1 - \frac{1}{\sqrt{G}} \tag{7.23}$$

With an operating (classical) OPO nonlinear gain of $G = 10.6$, the 'Classical Limit' curves in Fig. 7.7 show the region of classical noise spanned by deamplified and amplified forward seed power. The data trend falls within the 'Classical Limit' bounds.

To increase the precision of both measurements, the uncertainty in the input reverse-field amplitude, and the uncertainty in the relative phase between the forward field and the pump field needs to be reduced. To resolve the input amplitude issue, it is suggested that a mode-cleaning cavity be placed into the reverse-beam injection path, which will form a fixed-beam reference point. This will also potentially allow variation of the power without the need for optical filters, achieved with offsets to the cavity resonance condition. The control of the reverse-beam-to-pump relative phase is a trickier prospect. A potential solution is to introduce an piezo-electric transducer into the reverse-beam path to sweep the relative phase. Similar to the dual-scanning Local Oscillator phase sweep, this would average out the nonlinear amplification/deamplification effects.

### *7.6.2 Improving the Intrinsic Isolation*

The intrinsic isolation is dependent on all components of the OPO, not just the reflectivity of the nonlinear crystal. Recall that the DB-OPO cavity at ANU consists of non-superpolished mirrors (Sect. 6.2). Further, any form of scattering centre, such as poor optical coatings or the presence of dust particulates, will compromise the intrinsic isolation. Improved mirror/crystal surface polishing and optical coatings, as well as dust particulate minimisation, will see this isolation value increase.

## 7.7 Summary

The tolerance of a bow-tie OPO to backscattered light was modelled, and experimentally verified using the balanced homodyne detector. An intrinsic isolation of $(41 \pm 2)$ dB was measured, at least comparable to off-the-shelf Faraday isolators, and an additional suppression of approximately 40 dB over an equivalent standing-wave cavity. The BSDF of the OPO was determined to be $(1.7 \pm 0.3) \times 10^{-3}$, approximately ten times larger than the BRDF values of off-the-shelf reflective optics. Additional experiment hardware would give further precision in the measurements of the intrinsic isolation value, and improvements in the OPO optical components would further increase the isolation value.

## References

1. S.S.Y. Chua, M. Stefszky, C. Mow-Lowry, B. Buchler, S. Dwyer, D. Shaddock, P.K. Lam, D. McClelland, Backscatter tolerant squeezed light source for advanced gravitational-wave detectors. Opt. Lett. **36**(23), 4680 (2011)
2. H. Vahlbruch, S. Chelkowski, K. Danzmann, R. Schnabel, Quantum engineering of squeezed states for quantum communication and metrology. New J. Phys. **9**, 371 (2007)
3. M. Stefszky, C. Mow-Lowry, S. Chua, D. Shaddock, B. Buchler, H. Vahlbruch, A. Khalaidovski, R. Schnabel, P.K. Lam, D. McClelland, Balanced homodyne detection of optical quantum states at audio-band frequencies and below. Class. Quantum Gravity **29**, 145015 (2012)
4. The LIGO Scientific Collaboration, Nat. Phys. **7**, 962 (2011)
5. P. Horowitz, W. Hill, *The Art of Electronics*, 2nd edn. (Cambridge University Press, Cambridge, 1991)
6. Newport Corporation—Free-space Optical Faraday Isolators, http://www.newport.com/Faraday-Optical-Isolator-Free-Space/839201/1033/info.aspx
7. J.C. Stover, *Optical Scattering: Measurement and Analysis*, 3rd edn. (SPIE, Bellingham, 2012)
8. A.E. Siegman, *Lasers* (University Science Books, California, 1986)
9. P. Fritschel, Backscattering from the AS port: enhanced and advanced LIGO (2006). LIGO-T060303-00-D
10. M. Zucker, Comparison of video cameras for imaging diffuse scatter at 1064 nm wavelength (1999). LIGO-T990031-02-D
11. K. McKenzie, N. Grosse, W. Bowen, S. Whitcomb, M. Gray, D. McClelland, P.K. Lam, Squeezing in the audio gravitational-wave detection band. Phys. Rev. Lett. **93**(16), 161105 (2004)
12. K. McKenzie, Squeezing in the audio gravitational wave detection band. Ph.D. thesis, Physics Department, Australian National University, Canberra, 2008
13. K. Goda, K. McKenzie, Photothermal fluctuations as a fundamental limit to low-frequency squeezing in a degenerate optical parametric oscillator. Phys. Rev. A **72**, 043819 (2005)
14. T. Aoki, G. Takahashi, A. Furusawa, Squeezing at 946 nm with periodically poled $KTiOPO_4$. Opt. Express **14**, 6930–6935 (2006)

# Part III
# The LIGO H1 Squeezed Light Injection Experiment

# Chapter 8
# Overview of the LIGO Squeezed Light Injection Experiment

**Abstract** The LIGO Squeezed Light Injection Experiment was a project that first injected squeezed light into a LIGO 4 km interferometric gravitational-wave detector. This experiment was a collaboration between the LIGO Hanford Observatory, ANU Centre for Gravitational Physics, Massachusetts Institute of Technology, and the Max Planck Institute for Gravitational Physics (Albert Einstein Institute). The Experiment was co-led by the thesis author from ANU, Sheila Dwyer and Dr. Lisa Barsotti from MIT, and Dr. Daniel Sigg from LIGO Hanford Observatory, and aided by members of the LIGO Scientific Collaboration (LSC). After introducing the project background, an overview of the Enhanced LIGO interferometer is presented. This includes the interferometer hardware and modifications made for the LIGO Squeezed Light Injection Experiment. The squeezed light source for the Experiment is then covered, including specifics about the contribution from the ANU, the Optical Parametric Oscillator.

## 8.1 Project Background

The enhancement potential of quantum-noise limited gravitational-wave detectors using squeezed states was explored in Chap. 4. In addition to this potential capability, the reasoning to perform the Injection Experiment was threefold.

The first reason was to perform a squeezing injection targeting the lower detection frequencies below 300 Hz that future gravitational-wave detectors aim to measure.[1] The gravitational-wave frequency band below 300 Hz is an important region for astrophysical gravitational-wave sources [1]. This frequency region is susceptible to environmental and technical noise sources, such as backscattered light (Chap. 7). An unresolved area was whether an installed squeezer could potentially add noise to this important detection frequency region. The LIGO Squeezed Light Injection Experiment sought to answer this, and confirm the squeezing sensitivity-enhancement at the greater sensitivities and lower detection frequencies of a LIGO interferometer.

[1]By 2007, the preparations for injecting squeezing into the GEO600 detector were well underway. However, the GEO600 quantum-noise-limited sensitivity region is 700 Hz and above.



S.S.Y. Chua, *Quantum Enhancement of a 4 km Laser Interferometer Gravitational-Wave Detector*, Springer Theses,
DOI 10.1007/978-3-319-17686-4_8

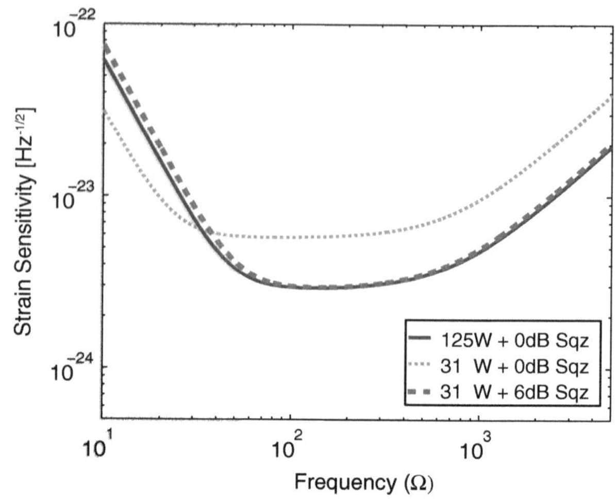


**Fig. 8.1** Example of injected squeezing recovering quantum-noise sensitivity at a lower interferometer optical power. This trace assumes a 6 dB squeezed state perfectly matched to the readout quadrature

The second reason was to test whether squeezing could be used as a safeguard for high interferometer power operation. Gravitational-wave detectors will deploy very high optical power levels within their Michelson arm cavities. For example, the Advanced LIGO detectors aim to have 800 kW of optical circulating power in each arm cavity [2]. At higher optical powers, the probability of wavefront distortion due to deformation of test masses via thermal absorption, and the probability of exciting parametric instabilities [3], are greatly increased.

Figure 8.1 shows an example on how injected squeezing can recover quantum-noise sensitivity with a lower interferometer optical power. An interferometer input laser power of 31 W with 6 dB of squeezing injected will nearly recover the quantum-noise-limited strain sensitivity of 125 W interferometer input power with no squeezing injection.[2] Thus injected squeezed states can be used to regain similar quantum noise performance with lower interferometer-operating laser powers if needed. The implicit additional benefit is with 125 W input power operation achieved, injected squeezing would further improve sensitivity without the need to inject even higher laser power. Therefore, squeezing provides both a way to improve interferometer sensitivity as well as an important safeguard for a technical risk source.

The third reason was to investigate techniques for the future installation/application of squeezed-light sources with gravitational-wave detectors. All third generation gravitational-wave detector designs utilize squeezed states (Chap. 2, Sect. 2.2.4). The LIGO Squeezed Light Injection Experiment, the first injection of squeezed light into a kilometre-scale gravitational-wave detector, would provide techniques, knowledge and experience in the implementation of squeezing in future gravitational-wave detectors.

[2]The difference being the effect of vacuum fluctuations from optical loss sources separate to the AS port vacuum fluctuations—see Appendix A for details.

A window of opportunity in 2011 to use the 4 km 'Hanford 1' or 'H1' Enhanced LIGO interferometer at the LIGO Hanford Observatory for such a squeezing experiment became available. This window worked around the Advanced LIGO upgrade schedule [4]. Beginning mid-2008, preparations began for the approved *LIGO Squeezed Light Injection Experiment.*
In summary, the reasoning and the goals of the Experiment were:

- To confirm the sensitivity-enhancement using injected squeezing at the greater strain-sensitivities and lower audio-detection frequencies of a LIGO interferometer.
- To investigate the interaction between the squeezed light source and interferometer. This included verifying that the presence of the squeezer did not cause additional noise on the interferometer readout.
- To investigate squeezer operation with LIGO, and techniques for the future installation/application of squeezed-light sources with gravitational-wave detectors.

## 8.2 The Enhanced LIGO Gravitational-Wave Detector, and Modifications for Squeezing Injection

The Enhanced LIGO (or eLIGO) gravitational-wave detector, and the hardware modifications made for the LIGO Squeezed-light Injection Experiment are described. This section will call upon the interferometer component definitions from Chap. 4.

### *8.2.1 The Enhanced LIGO H1 Detector*

The eLIGO H1 interferometer had the greatest (unsqueezed) strain sensitivity to gravitational waves yet achieved [5]. A simplified diagram of the eLIGO H1 interferometer is shown in Fig. 8.2. The interferometer's light source is a Nd:YAG laser, producing 35 W of 1064 nm light. This laser is stabilised in laser frequency and laser intensity noise, hence the name of *Pre-Stabilised Laser* or the *PSL*. The PSL light passes through an electro-optic modulator (EOM) where the *interferometer control sidebands* are phase modulated onto the light. After spatial-mode filtering by the interferometer's *input mode cleaner* (*IMC*), the light is then incident on the interferometer. The interferometer is a Power-recycled Fabry-Perot Michelson, the configuration of Sect. 4.4. Table 4.1 lists the key parameters for the eLIGO H1 interferometer. The core interferometer optical components (the PRM, BS, ITMs and ETMs—Chap. 4) are all suspended from vibration-isolation platforms to reduce the coupling of seismic motion to the optics.

The interferometer control sidebands are transmitted to the AS port by design. The Michelson is also controlled so there is a small offset from total destructive interference at the AS port. With the *dark fringe offset*, a small amount of light is

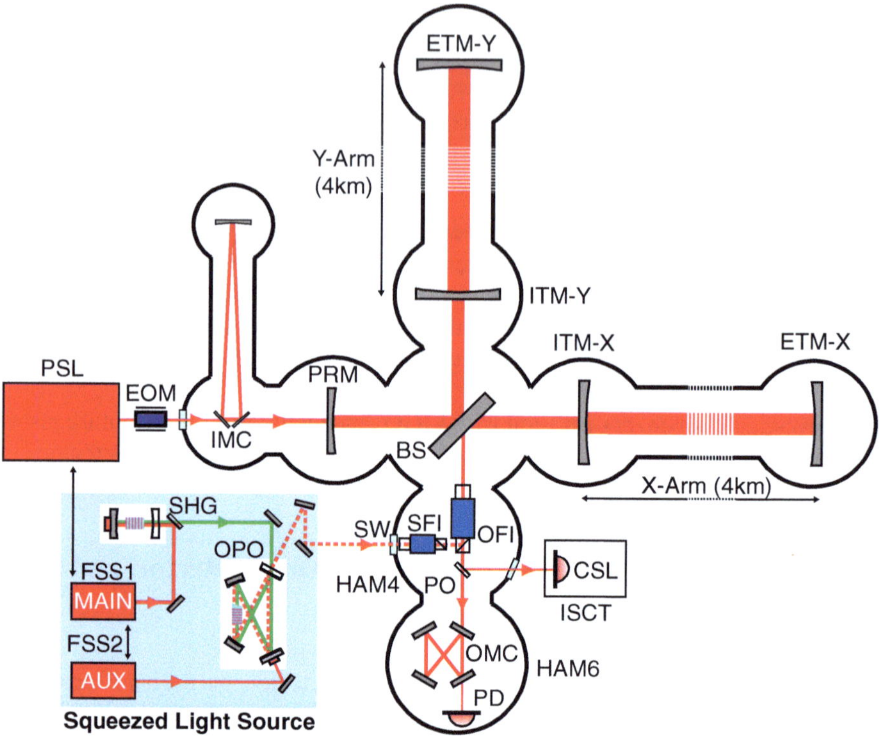


**Interferometer Acronyms**

| | |
|---|---|
| PSL | - Pre-Stabilized Laser |
| — | - Vacuum System |
| IMC | - Input Mode Cleaner Cavity4 |
| EOM | - Electro-Optic Modulator (Control Sidebands) |
| PRM | - Power Recycling Mirror |
| BS | - Beam Splitter |
| ITM | - Arm Cavity Input Test Mass |
| ETM | - Arm Cavity End Test Mass |
| HAM | - Horizontal Acess Module Vacuum Chamber |
| OFI | - Faraday Isolator |
| PO | - Pick Off Mirror |
| ISCT | - Interferometer Sensing and Control Table |
| OMC | - Output Mode Cleaner Cavity |
| PD | - Readout Photodetector |

**Squeezed Light Source Acronyms**

| | |
|---|---|
| MAIN | - Squeezing Main Laser |
| AUX | - Squeezing Auxiliary Laser |
| FSS | - Frequency Stabilisation Servo |
| SHG | - Second Harmonic Generator |
| OPO | - Optical Parametric Oscillator |
| SFI | - Squeezing Faraday Isolator |
| SW | - Septum Window |
| CSL | - Coherent Locking Photodetector |

**Fig. 8.2** Simplified schematic of the Enhanced LIGO interferometer, along with the location of the squeezer and the additional hardware modifications made for the injection experiment. Components not drawn to scale. Key components only shown for clarity

transmitted to the AS port, the *carrier field*. The optical field composition at the AS port sees the largest proportion (at least 80 %) of the optical power contained within the interferometer control sidebands.

The AS port beams then pass through the Output Faraday Isolator (OFI), to prevent reflected scatter from optics in the readout chain from re-entering the interferometer.

After the OFI, a small amount of the power at the antisymmetric port is tapped off (with the Pick Off mirror—PO) and sent to the *Interferometer Sensing and Control Table (ISCT)*. This optical table, situated outside the vacuum system, contains several photodetectors used for interferometer alignment sensing and control, including the sensors that measure the interferometer control sidebands. The remaining AS port power is then incident on the Output Mode Cleaner (OMC). The OMC spatially-filters the AS port beam to create a quiet Local Oscillator for homodyne detection (called DC Readout [6, 7]) at the interferometer readout photodetector. This filters out the remaining interferometer control sidebands, coherent locking sidebands, and any higher-order spatial modes, leaving a spatially-filtered gravitational-wave signal and carrier Local Oscillator beam. The OMC is discussed further in Appendix B

### 8.2.2 Hardware Installed for Squeezing Injection

In addition to the squeezed light source, several hardware components were installed for the LIGO Squeezed Light Injection Experiment. These components are also shown in Fig. 8.2.

- In-vacuum component installations in the AS port—The original eLIGO OFI was replaced with a new Advanced LIGO OFI. This modification was needed as the Advanced LIGO OFI has an open port polariser available for squeezing injection. The role of the open port is discussed in the next section (8.2.3). An additional Faraday isolator, known as the Squeezing Faraday Isolator (SFI) was installed in the squeezing injection path for additional attenuation of beams from the interferometer that were incident on the squeezer table.
- On the ISCT, the Coherent Sideband locking photodetector (CSL), used for reading out the relative phase between the coherent locking sideband and the AS port carrier Local Oscillator, was installed.
- The Output Mode Cleaner of the H1 interferometer was swapped with the Output Mode Cleaner of the LIGO L1 interferometer. This was necessary due to the sudden increase in optical loss by the original OMC installed at Hanford [7]. Details about the reasoning for the swap and measurements made to determine the increase in optical loss, are presented in Appendix B.

### 8.2.3 Squeezed-Light Injection Path

Further shown in Fig. 8.2 is the squeezed light injection path. The squeezed light (and co-propagating coherent locking field) are injected into the interferometer vacuum system via the septum window (SW). After passing the SFI, the fields are then injected into the OFI which acts like an *optical circulator*. Mentioned in Chap. 4, the optical circulator allows the injection of squeezing without additional loss to

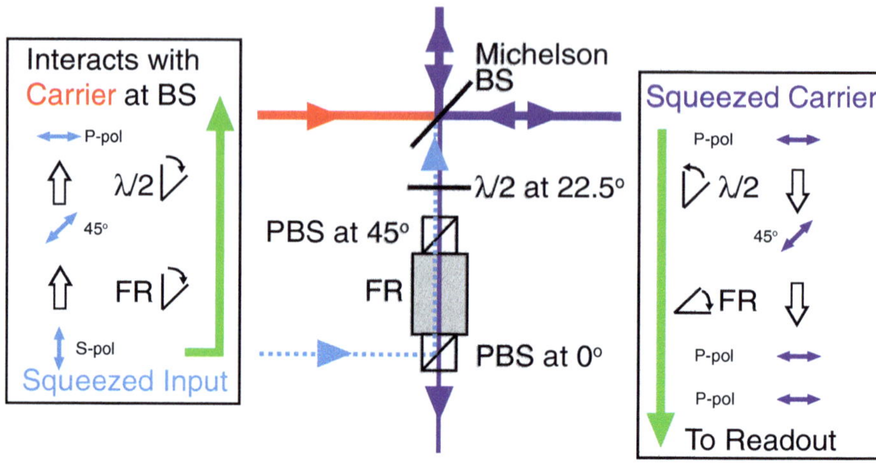


**Fig. 8.3** Squeezing injection using a Faraday isolator as an optical circulator. The polarisations are illustrated in the panels either side of the schematic. *S-pol* s-polarisation, *P-pol* p-polarisation, *FR* Faraday rotator, *PBS* polarising beamsplitter, $\lambda/2$ half-wave plate

the interferometer field. Figure 8.3 illustrates squeezing being injected into an open polariser port of the isolator, and the field polarisations allowing a Faraday isolator to become an optical circulator. The open polariser port requirement necessitated the installation of an Advanced LIGO OFI. Faraday isolators have previously been used as optical circulators in squeezing-interferometer proof-of-principle experiments (see Sect. 1.1.3).

The squeezing (and coherent locking sidebands) then become incident on the interferometer beamsplitter, to then reflect and co-propagate with the carrier field. After the OFI, some of the AS port optical light is the tapped off with the PO and is directed to the ISCT, where the sensing photodetector for the LO coherent locking loop is located. The SW, SFI, OFI and PO are all part of/ housed in a vacuum chamber called the *Horizontal Access Module* (*HAM*) number 4, thus this location will be referred to as *HAM4*. The remaining AS port power is then incident on the OMC. The OMC and interferometer readout photodetector are housed in a separate vacuum chamber named *HAM6*.

### *8.2.4 Sensitivity of the Enhanced LIGO H1 Interferometer*

Figure 8.4 shows the interferometer readout spectrum, and the sources of limiting noise. The mechanisms for these noise sources were presented in Chap. 2. The interferometer readout spectrum represents the measurement sensitivity of the interferometric-detector to gravitational waves. There are several important features contained in Fig. 8.4.

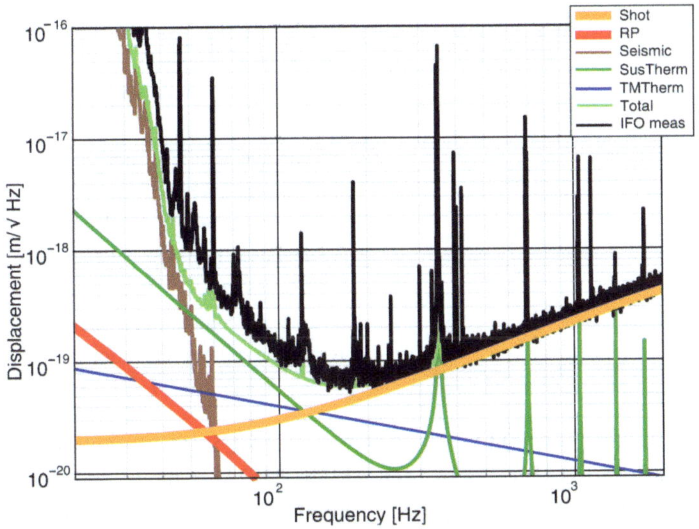


**Fig. 8.4** The eLIGO sensitivity curve and various noise levels contributions—*Shot* quantum shot noise, *RP* quantum radiation pressure noise, *Seismic* seismic ground motion, *SusTherm* mirror suspension thermal noise, *TMTherm* test mass thermal noise, *Total* sum of contributing noise sources, *IFO meas* measured interferometer sensitivity

- The region that is quantum shot-noise limited is approximately 200 Hz and above, while thermal noise of the test mass suspensions and seismic noise, become dominant at frequencies starting from 150 Hz and below. Squeezing improvement should be observable above 200 Hz, and steadily degraded from 200 Hz and below.
- Due to the dominance of seismic and suspension thermal noise, the quantum radiation-pressure regime is completely masked by at least an order of magnitude, thus quantum-radiation-pressure phenomena observation is unobtainable.
- The units of this spectrum are in terms of *equivalent test mass displacement*, $\Delta L$. This is related to the strain via the arm-cavity length (Eq. 2.1). For the awareness of the reader, the interferometer readout spectrum will be presented in a number of different units in subsequent chapters.

## 8.3 The LIGO H1 Squeezed Light Source

The Injection Experiment's squeezed-light source, the *LIGO H1 Squeezer*, was a new squeezer, with all partner institutions delivered components of the apparatus. These contributions are listed in Table 8.1. Figure 8.5 shows the layout of the LIGO H1 squeezer. The functions correspond closely with the 'generic' squeezer components of Chap. 5.

**Table 8.1** LIGO squeezer component contributions from the partner institutions

| Partner institution | Squeezer component |
|---|---|
| ANU | o OPO testing/commissioning<br>o OPO delivery |
| MIT | o Main and auxiliary lasers<br>o SHG construction<br>o Squeezer pre-assembly laboratory |
| AEI | o Diagnostic homodyne detector<br>o SHG design |
| LIGO | o Squeezer control electronics<br>o Interferometer preparation |

The two on-table lasers (MAIN and AUX) were Nd:YAG NPRO lasers. Two laser-frequency stabilisation servo loops (denoted by FSS) kept the Main laser frequency-locked to the PSL frequency (loop FSS1) and the Auxiliary laser locked to the Main laser with a 29.5 MHz offset (loop FSS2). These were measured with the photodetectors labelled 'FSS1 PD' and 'FSS2 PD' respectively. The LIGO H1 squeezer used the unmodified Coherent Sideband Locking scheme (Sect. 6.3.1), hence the presence of the second frequency stabilisation loop.

The Main laser was used to drive the Second Harmonic Generator (SHG), generating the pump field at 532 nm. The SHG was a standing-wave cavity with a PPKTP crystal, mounted in a temperature-controlled peltier oven. The cavity length of the SHG was controlled using the PDH technique on the Main laser field, measured in transmission on 'SHG PD' with feedback to a PZT-mounted cavity mirror.

The SHG produced the pump beam used to drive the Optical Parametric Oscillator (OPO). The OPO cavity length was controlled using PDH locking with the reflected pump beam, measured by the 'OPO Refl' photodetector. The squeezed beam produced by the OPO is separated from the reflected green beam by a dichroic mirror and sent either to the diagnostic homodyne detector or into the interferometer. The homodyne detector was the same homodyne that arrived at the ANU for the LIGO H1 OPO testing period (see Sect. 6.6).

The LO field for the diagnostic homodyne detector was a tap-off beam from the PSL, spatially-mode-cleaned using the transporting single-mode optical fibre (Sect. 5.5.1). The installed Mach-Zehnder interferometer for pump power stabilisation (Sect. 5.5.2) was not used in the Experiment. After the 'Squeezing Flipper Mirror' (used to direct the produced squeezing to the diagnostic homodyne detector), the half-wave plate (HWP) was used to adjust the squeezing polarisation, and two remote-controllable PZT steering mirrors allowed for squeezing beam alignment. The large range PZT was not used in squeezing-enhancement measurements, but for backscattering measurements reported in Chap. 10.

The unmodified coherent sideband locking technique (Sect. 6.3.1) was implemented for squeezing angle control. The Auxiliary laser field was injected into the second flat mirror of the OPO. The pump-auxiliary relative phase was measured in

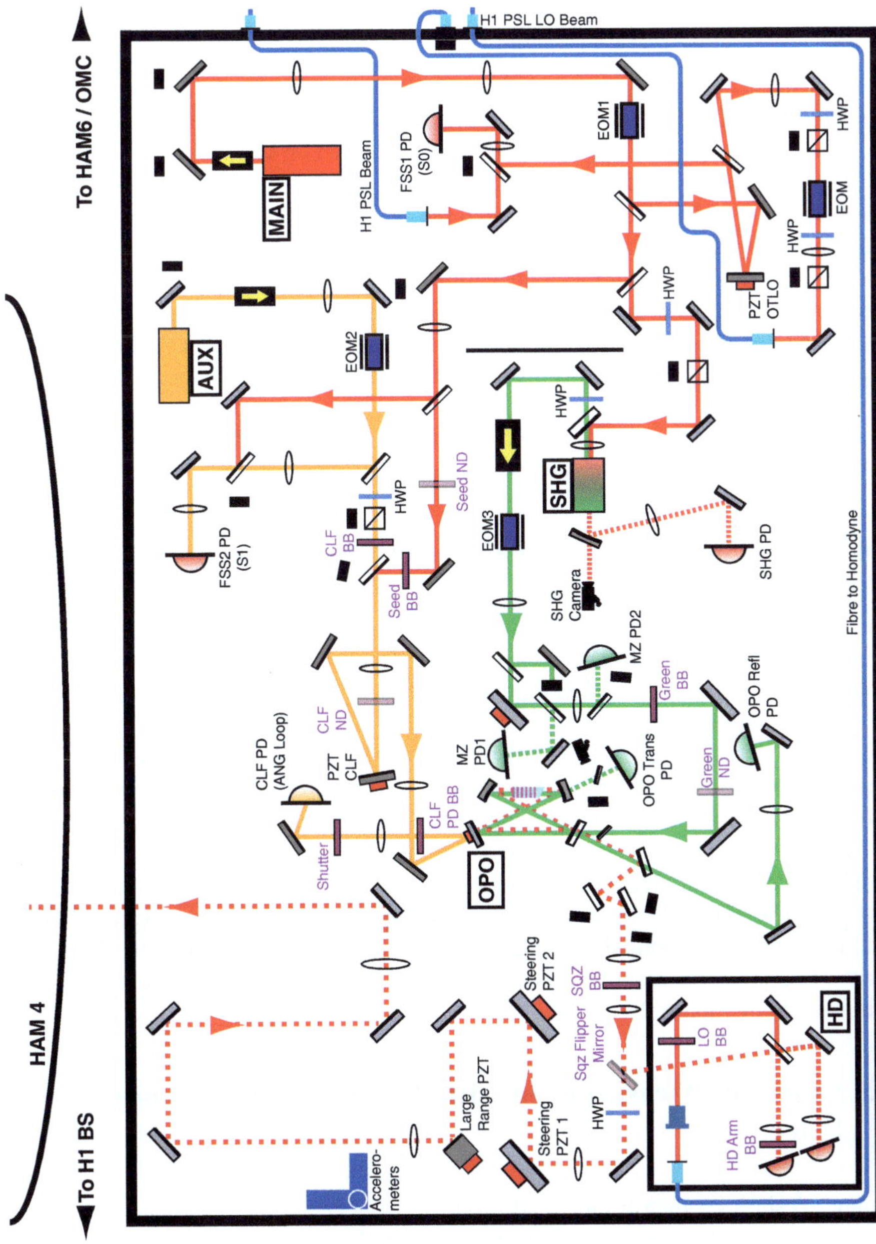


**Fig. 8.5** Schematic of the LIGO H1 squeezer. The key components correspond in function to the 'generic squeezer' of Chap. 5, and are detailed in Sect. 8.3. Optical lens values are omitted for clarity. The optics labelled 'BB' are optical beam blocks and 'ND' are neutral-density filters for beam power attenuation. Beam blocks and neutral density filter were all mounted on automated flipper mounts to allow testing of the squeezer remotely. The on-table accelerometers measured the squeezing table motion, used in measurements of backscatter (Chap. 10)

reflection by the 'CLF PD' photodetector and was actuated using the PZT in the auxiliary beam path ('PZT CLF'). The second coherent sideband locking control loop was measured by the diagnostic homodyne detector (when used), or a photodetector located on ISCT. When operating with the interferometer, the actuator to control the squeezing ellipse phase (and to close the 'LO' loop) was a Voltage Control Oscillator (VCO). The VCO added a correcting-frequency offset to the PSL-Main frequency control loop (FSS1), that then controlled the squeezing phase via the cascade of the Main, SHG and OPO control loops.

The electronics used for the control loops leveraged the designs of analog control electronics being used in Advanced LIGO. These included analog electronic frequency synthesizers, RF distribution panels, and RF demodulator boards, all with digital gains, offsets and polarity switches. Remote controls and user-interfaces were achieved using the Beckhoff TWINCAT system [8], interfaced with the interferometer computing system (EPICS—[9, 10]) at the Observatory.[3] This allowed operation of the squeezer from the interferometer main control room.

## 8.4 The LIGO H1 Optical Parametric Oscillator

The Optical Parametric Oscillator was the component designed and delivered by the ANU Centre for Gravitational Physics. The LIGO H1 Optical Parametric Oscillator cavity design closely resembled the DB-OPO (Sect. 6.2), namely a doubly-resonant travelling-wave cavity in bow-tie configuration with a wedged PPKTP nonlinear crystal. For the remainder of this thesis, this OPO will known as the 'H1 OPO'. The H1 OPO was tested at the ANU in the January 2010 period, where greater than 5 dB of squeezing was measured from 100 Hz and above, and greater than 6 dB from 200 Hz and above (see Sect. 6.6—Fig. 6.13).

There was an important difference between the H1 OPO and the final DB-OPO at ANU. The H1 OPO cavity mirrors were super-polished mirrors from ATFilms [11] with equivalent optical-coating reflectivities. This difference will be significant for the backscatter measurements in Chap. 10. The H1 OPO cavity parameters are summarised in Table 8.2.

## 8.5 Squeezing Measurement with the Diagnostic Homodyne Detector

Figure 8.6 shows a measurement of produced squeezing and antisqueezing on the diagnostic homodyne detector. This was measured on a Stanford Research Systems SR785 Spectrum Analyzer. The input pump power to the OPO was 50 mW, the operating pump power for the LIGO H1 Squeezer.

[3] The author acknowledges, with gratitude, the remote controls and user-interfaces were coded and maintained by Maxim Factourovich from Columbia University NY, under the supervision of Dr. Daniel Sigg.

**Table 8.2** LIGO H1 squeezer optical parametric oscillator parameters

| Cavity parameter | Symbol | Value | Units |
|---|---|---|---|
| Fundamental wavelength | $\lambda_a$ | 1064 | nm |
| Second harmonic (pump) wavelength | $\lambda_b$ | 532 | nm |
| Round-trip optical path length | L | 0.279 | m |
| Curved mirrors' radius of curvature | RoC | −38 | mm |
| Input/output coupler reflectivity (fundamental) | $R^a_{in/out}$ | 0.868 | – |
| Input/output coupler reflectivity (pump) | $R^b_{in/out}$ | 0.696 | – |
| PZT mirror reflectivity | $R^a_2$ | 0.998 | – |
| Total intra-cavity loss at fundamental | $T^a_l$ | 0.0056 | – |
| Total intra-cavity loss at pump | $T^b_l$ | 0.039 | – |
| Finesse (fundamental) | $\mathcal{F}_a$ | 42.5 | – |
| Finesse (pump) | $\mathcal{F}_b$ | 15.6 | – |
| Full-width-half-maximum (fundamental) | $\Delta\nu_a$ | 25.8 | MHz |
| Full-width-half-maximum (pump) | $\Delta\nu_b$ | 70.0 | MHz |
| Escape efficiency | $\eta_{esc-H1}$ | 0.959 | – |
| Threshold power | $P_{thr-H1}$ | 75 | mW |

A squeezing magnitude of $6.2 \pm 0.2$ dB and antisqueezing magnitude of $10.6 \pm 0.2$ dB between 9 and 12 kHz was observed. The squeezing magnitude measured is smaller compared to the magnitude of squeezing from the DB-OPO (Sect. 6.8.4). This is due to a reduced squeezing detection efficiency of $\eta_{\text{det}} = 85 \pm 4\,\%$. The features below 3 kHz were due to backscattered-light interferences that are sourced from the diagnostic homodyne local oscillator beam. The feature at 5 kHz was caused by an oscillation of the pump-to-coherent sideband control loop.

Due to constraints on experiment time, the diagnostic homodyne detector was not further optimised. With optimisation of the detection efficiency and the mitigation of backscattered light, the measurement would improve in terms of both squeezing magnitude and lower Fourier frequencies. The backscattered-light interferences of the homodyne do not impact the interferometer measurement, as the diagnostic homodyne is bypassed entirely. The 5 kHz oscillation was nullified by adjustment of control electronic gains.

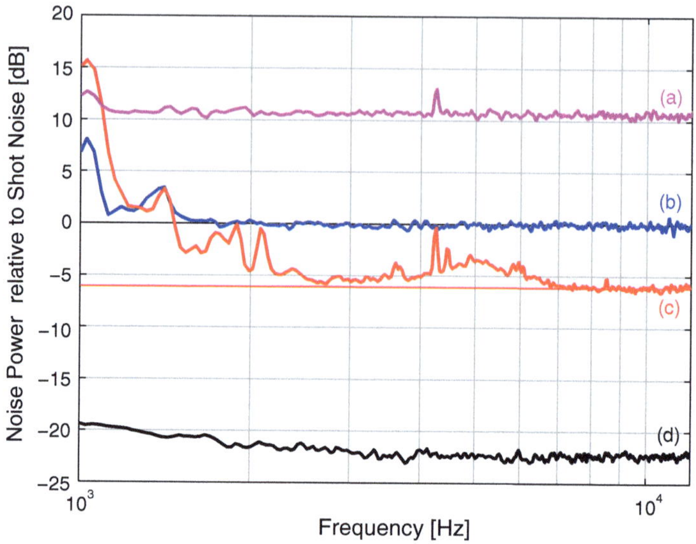


**Fig. 8.6** Measurement of squeezing with the diagnostic homodyne detector. Traces are **a** anti-squeezing, **b** shot noise, **c** squeezing and **d** electronic dark noise. A *line* shows the average squeezing level between 9 to 12 kHz, measured at 6.2 ± 0.2 dB. Antisqueezing level is at 10.6 ± 0.2 dB above shot noise. All traces are measured with a single span 0–25.2 kHz with 800 FFT lines (resolution bandwidth of 32 Hz), and taken with 300 rms averages. Dark noise has not been subtracted from the other traces

## 8.6 Summary

This chapter sets the background for the LIGO Squeezed-light Injection Experiment. The project goals and the Enhanced LIGO interferometer were introduced, with the hardware modifications made for this experiment and the squeezed light injection optical path. The noise sources limiting the detector sensitivity were presented, particularly showing the quantum shot-noise limited region to be 200 Hz and above, and seismic and suspension thermal noise becoming, then dominating below 200 Hz. The squeezed-light source and the ANU-contributed H1 OPO were presented. This chapter described the setting and setup on which the results presented in the next chapters will be based.

## References

1. C. Cutler, K.S. Thorne, An overview of gravitational-waves sources (2002), arXiv:gr-qc/0204090v1
2. G.M. Harry (for the LIGO Scientific Collaboration), Advanced LIGO: the next generation of gravitational wave detectors. Class. Quantum Gravity **27**, 084006 (2010)

3. J.A. Sidles, D. Sigg, Optical torques in suspended Fabry Perot interferometers. Phys. Lett. A **354**(167) (2006)
4. D. Sigg, N. Mavalvala, D. McClelland, P.K. Lam, R. Schnabel, H. Vahlbruch, S. Whitcomb, Proposal for a squeezed H1 interferometer (2007). LIGO-T070265-D-D
5. The LIGO Scientific Collaboration and The Virgo Collaboration (J. Abadie et. al.), Sensitivity achieved by the LIGO and virgo gravitational wave detectors during LIGO's sixth and virgo's second and third science runs (2012), arXiv:1203.2674v2
6. T.T. Fricke, N.D. Smith-Lefebvre, R. Abbott, R. Adhikari, K.L. Dooley, M. Evans, et al., DC readout experiment in Enhanced LIGO. Class. Quantum Gravity **29**(6), 065005 (2012)
7. T. Fricke, Homodyne detection for interferometric gravitational-wave detectors. Ph.D. thesis, Graduate Faculty, Louisiana State University, 2011
8. Beckhoff Automation LLC (TWINCAT), http://www.beckhoff.com/
9. Experimental Physics and Industrial Control System (EPICS), http://www.aps.anl.gov/epics/index.php
10. EPICS Motif Editor and Display Manager (MEDM), http://www.aps.anl.gov/epics/extensions/medm/index.php
11. ATFilms, http://www.atfilms.com/

# Chapter 9
# Squeezing-Enhancement of a 4 km LIGO Gravitational-Wave Detector

**Abstract** This chapter presents the main results of the LIGO Squeezed-light Injection Experiment—the sensitivity-enhancement and characterisation of the Enhanced LIGO interferometer with injected squeezed states. These results represent the first measurement of squeezing enhancement in a 4 km interferometric gravitational-wave detector, and the best high-frequency sensitivity in a gravitational-wave detector yet achieved. The investigations undertaken to characterise the squeezing injection parameters are first reported. The results of measurements of squeezing ellipse phase noise are presented in Sect. 9.1, followed by measurements made to characterise the squeezing injection detection efficiency in Sect. 9.2.

The squeezing enhancement results are then presented:

- The squeezing-enhanced LIGO interferometer spectrum is shown in Sect. 9.3, along with a squeezing-improvement calculation, and the highlighting of the 150 Hz detection frequency region.
- The squeezing-enhanced LIGO spectrum is then compared with the sensitivity-data from the joint scientific run in Sect. 9.4, showing the squeezing-enhanced spectrum to be best high-frequency sensitivity in a gravitational-wave detector yet achieved.
- The interferometer operating power is theoretically scaled in Sect. 9.5 to match the squeezing-enhanced quantum noise spectrum. This is to demonstrate the regaining of similar quantum noise performance with lower interferometer-operating laser powers.
- In Sect. 9.6, the squeezing magnitude generated from the OPO is inferred, then forward-calculated for the expect squeezing-improvement in the interferometer sensitivity. This is calculated from the squeezing measurement efficiency and phase noise results.

Recommendations for improving the squeezing enhancement in future detectors conclude the Chapter (Sect. 9.7).



S.S.Y. Chua, *Quantum Enhancement of a 4 km Laser Interferometer Gravitational-Wave Detector*, Springer Theses,
DOI 10.1007/978-3-319-17686-4_9

## 9.1 Squeezing Ellipse Phase Noise

As previously detailed in Chap. 6 (Sect. 6.4.4), squeezing ellipse phase noise can mix the antisqueezed quadrature noise into squeezing measurement, reducing the squeezing improvement. This section presents results of measurements of phase noise originating from:

- the squeezed light source
- interferometer control sidebands interacting with the interferometer contrast defect
- interferometer control sidebands unbalancing or asymmetry, and
- beam pointing fluctuations

### *9.1.1 Phase Noise from the LIGO Squeezed Light Source*

The measurement of phase noise from the LIGO squeezed light source was made with a squeezing/antisqueezing measurement as a function of OPO nonlinear gain. This is the same test made to measure the phase noise from the ANU squeezed light source in Chap. 6 (Sect. 6.4.5).

With the diagnostic homodyne detector, spectra of shot noise, locked squeezing and locked anti-squeezing were measured over different values of OPO operating nonlinear gain. These spectra were recorded with an Agilent SR785 Spectrum Analyzer. The unity gains of the control loops for the OPO cavity length and coherent locking were kept approximately constant by adjusting the electronic gain on the analog boards. The homodyne photodetector efficiency was 91 $\pm$ 2 %. The propagation efficiency was 98 $\pm$ 1 %—the dominant propagation loss component was the flipper mirror used to direct the squeezed beam towards the diagnostic homodyne (thus was not part of the interferometer injection path). The homodyne efficiency was measured to be 95 $\pm$ 1 %.

The measurement is shown in Fig. 9.1. The total measurement efficiency is 81 $\pm$ 4 % (Eq. 6.15), consistent with the product of the photodiode efficiency, propagation efficiency, homodyne efficiency and OPO escape efficiency. The phase noise fitted to the data, and hence the inferred rms-phase noise contribution from the squeezed light source is $\tilde{\theta}_{\mathrm{sqzr}} = 22 \pm 3$ mrad.

### *9.1.2 Phase Noise Due to the Interferometer Control Sidebands*

Phase noise of the optical beams in the interferometer AS port directly contributes to the total phase noise measured at the readout photodetector. This noise causes fluctuations of the readout quadrature. As shown in Fig. 9.2, for a fixed squeezing ellipse angle, a fluctuating readout quadrature couples in the antisqueezing noise.

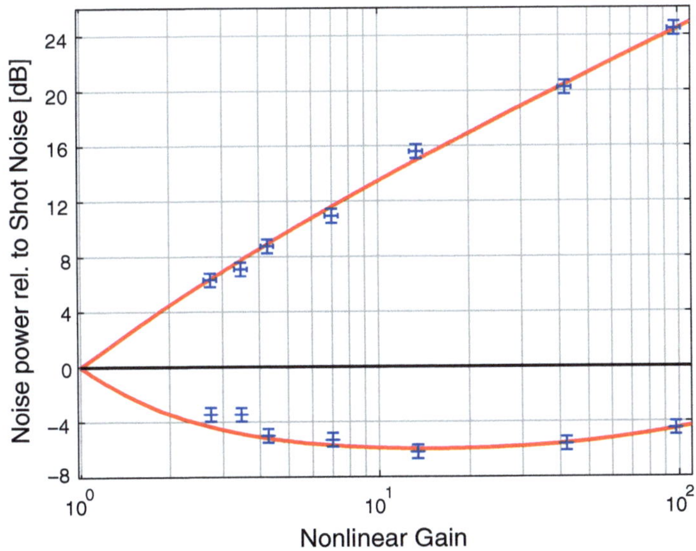


**Fig. 9.1** Phase noise measurement of the squeezer. Data measured on the diagnostic homodyne detector on the squeezer optical table. *Fitted curves* for total measurement efficiency of $\eta_{tot} = 81 \pm 4\,\%$ and rms phase noise of $\tilde{\theta}_{\text{sqzr}} = 22 \pm 3$ mrad

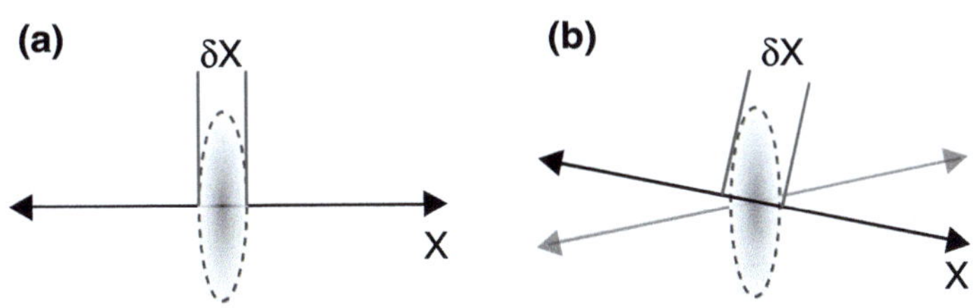


**Fig. 9.2** The effect of readout phase noise on the vacuum squeezing measurement. **a** No readout phase noise, thus the measurement quadrature (X) measures the true squeezing magnitude ($V_X = \langle \delta X \rangle^2$). **b** Readout phase noise present, thus the readout quadrature is pivoting, resulting in the coupling of the antisqueezed quadrature noise. The measurement no longer reflects the true squeezing magnitude

The measurement no longer reflects the true squeezing magnitude. This is equivalent to the effect of a fixed readout quadrature and a fluctuating squeezing ellipse angle (comparing Fig. 9.2 to Fig. 6.10).

Figure 9.3 illustrates the phasor diagram progression of the main fields in the interferometer. The phasors in Fig. 9.3 correspond to the variables listed in Table 9.1. We begin with the input beam to the Michelson [Fig. 9.3 panel (a)], consisting of the carrier field and the interferometer phase modulation sidebands used for controlling the Michelson. After interacting with the Michelson, the main optical fields at the AS port are [1]:

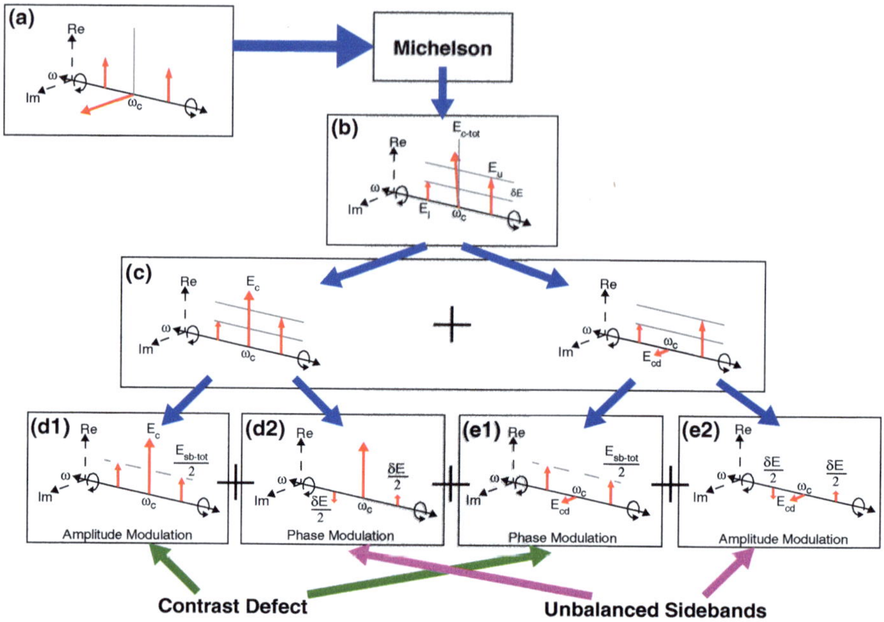


**Fig. 9.3** Phasor diagrams of the interferometer output. Labels correspond to fields listed in Table 9.1 and discussed in Sect. 9.1.2. **a** The input field, showing the carrier and interferometer control phase modulation (PM) sidebands. **b** The Michelson rotation of the carrier, and the contrast defect field and interferometer sidebands (unbalanced with exaggeration), unchanged by the Michelson. **c** Decomposition of the AS port field into carrier and contrast defect fields. **d1** The effect of interferometer control sidebands causing amplitude modulation (AM) on the carrier field. **d2** The effect of unbalanced sidebands on the carrier field, causing PM of the carrier field. **e1** The effect of interferometer control sidebands causing PM on the contrast defect field. **e2** The effect of unbalanced sidebands on the contrast defect field, causing AM

**Table 9.1** Variables used in calculations of phase noise from interferometer control sidebands

| Variable | Symbol |
|---|---|
| Carrier field amplitude | $E_c = \sqrt{P_c}$ |
| Contrast defect field amplitude | $E_{cd} = \sqrt{P_{cd}}$ |
| Total readout field amplitude | $E_{c-tot} = \sqrt{P_{tot}} = \sqrt{P_c + P_{cd}}$ |
| Upper sideband field amplitude | $E_u = \sqrt{P_u}$ |
| Lower sideband field amplitude | $E_l = \sqrt{P_l}$ |
| Total sideband field amplitude | $E_{sb-tot} = \sqrt{2\bar{P}_{sb}} = \sqrt{\lvert P_u + P_l \rvert}$ |
| Sideband difference amplitude | $\delta E = \sqrt{(\delta P)} = \sqrt{\lvert P_u - P_l \rvert}$ |
| Sideband (power) OMC transmission | $T_{sb}$ |
| Sideband modulation frequency | $\omega_{sb}$ |

The variables are assigned with their phasors in Fig. 9.3

- The carrier field—The optical field used as the LO field for reading out the gravitational-wave signal.
- The interferometer control sidebands—Transmitted interferometer sidebands to the AS port, by design.
- The contrast defect field—An optical field at the carrier frequency, a result of the asymmetry of the Michelson arm reflectivities. This asymmetry is caused by differences in both reflectivity magnitude and reflection beam spatial overlap. This field is measurable when the dark fringe offset of the interferometer is set to zero.

The output carrier field phase is rotated by 90° by the Michelson, but the other fields remain unrotated. Further, in association with the Michelson operating state, each of the interferometer control sidebands may have different optical powers in the AS port.

Before reaching the interferometer readout photodetector, the Output Mode Cleaner (resonant on the carrier frequency) attenuates most of the control sideband power. However, a small proportion of sideband power reaches the readout photodetector due to the finite cavity finesse. The Output Mode Cleaner being on-resonance for the carrier field frequency means that it is also resonant for the contrast defect field, thus there is only spatial-mode attenuation of the contrast defect field.

The phasors at the readout photodetector [panel (b)] can be rearrange to account for the *contrast defect* and the *sideband unbalancing* phase fluctuations [panels (c), (d1,d2) and (e1,e2)] due to the interferometer sidebands. Definitions of the variables used to calculate these phase noise contributions are provided in Table 9.1, corresponding with the labelled phasors in Fig. 9.3. The readout quadrature phase $\phi_{\rm rd}$ due to the presence of the contrast defect field is described by

$$\begin{aligned}\phi_{\rm rd} &\approx \tan(\phi_{\rm rd})\\ &= \frac{E_{\rm cd}}{E_{\rm c}}\end{aligned}\tag{9.1}$$

This is the phase the squeezing ellipse phase must match for maximum squeezing enhancement. The readout quadrature phase angle is determined by the carrier and the contrast defect fields, therefore fluctuations in either field will be a source of phase noise. The next subsections calculate the phase noise contributions from the contrast defect and unbalanced interferometer control sidebands.

#### 9.1.2.1 Contrast Defect Phase Fluctuations

The presence of the interferometer control sidebands causes both *amplitude modulation (AM) of carrier* [panel (d1)] and *phase modulation (PM) of the contrast defect* [panel (e1)] fields. These are sources of phase fluctuations of the readout quadrature. We call upon the classical modulation equations of Chap. 3 (Sect. 3.3.1, Eqs. 3.3.2 and 3.3.4) to calculate these contributions.

Firstly, the modulation depth $M_{\mathrm{cd}}$ is given by the proportion of total sideband field relative to the total readout field

$$M_{\mathrm{cd}} = \frac{E_{\mathrm{sb-tot}}}{E_{\mathrm{c-tot}}} = \sqrt{\frac{T_{\mathrm{sb}} \times 2\bar{P}_{\mathrm{sb}}}{P_{\mathrm{c-tot}}}} \tag{9.2}$$

Using Eq. 3.3.2, the amplitude modulated carrier, shown by Fig. 9.3d1, is written as

$$E_{\mathrm{c-AM}} = E_{\mathrm{c}}(1 + M_{\mathrm{cd}} \cos(\omega_{\mathrm{sb}} t)) \tag{9.3}$$

The phase of the readout quadrature is now

$$\begin{aligned} \phi_{\mathrm{rd}} &= \frac{E_{\mathrm{cd}}}{E_{\mathrm{c-AM}}} \\ &= \frac{E_{\mathrm{cd}}}{E_{\mathrm{c}}(1 + M_{\mathrm{cd}} \cos(\omega_{\mathrm{sb}} t))} \\ &\approx \frac{E_{\mathrm{cd}}}{E_{\mathrm{c}}}(1 + M_{\mathrm{cd}} \cos(\omega_{\mathrm{sb}} t)) \end{aligned} \tag{9.4}$$

where the last line was reached using a first-order Taylor approximation ($M_{\mathrm{cd}} \ll 1$) and noting that the cosine function is an even function. Compared with Eq. 9.1, the change in phase is

$$\Delta\phi_{\mathrm{c-AM}} = \frac{E_{\mathrm{cd}}}{E_{\mathrm{c}}} M_{\mathrm{cd}} \cos(\omega_{\mathrm{sb}} t) \tag{9.5}$$

Using $\overline{\cos^2(x)} = 1/2$, the rms-phase fluctuation due to the AM carrier is therefore

$$\begin{aligned} \tilde{\theta}_{\mathrm{c-AM}} &= \sqrt{\overline{|\Delta\phi_{\mathrm{c-AM}}|^2}} \\ &= \sqrt{\frac{P_{\mathrm{cd}}}{P_{\mathrm{c}}} M_{\mathrm{cd}}^2 \times \frac{1}{2}} \\ &= \sqrt{\frac{1}{2} \frac{T_{\mathrm{sb}} 2\bar{P}_{\mathrm{sb}} P_{\mathrm{cd}}}{P_{\mathrm{c}} P_{\mathrm{c-tot}}}} \end{aligned} \tag{9.6}$$

For the phase-modulated contrast defect field shown by Fig. 9.3e1, using Eq. 3.34

$$E_{\mathrm{cd-PM}} = E_{\mathrm{cd}}(1 + i M_{\mathrm{cd}} \cos(\omega_{\mathrm{sb}} t)) \tag{9.7}$$

$$\Delta\phi_{\mathrm{cd-PM}} = \frac{E_{\mathrm{cd}}(i M_{\mathrm{cd}} \cos(\omega_{\mathrm{sb}} t))}{E_{\mathrm{c}}} \tag{9.8}$$

The rms-phase fluctuations due to the phase modulated contrast defect field is

$$\begin{aligned}\tilde{\theta}_{\text{cd}-\text{PM}} &= \sqrt{\overline{|\Delta\phi_{\text{cd}-\text{PM}}|^2}} = \sqrt{\overline{(\Delta\phi_{\text{cd}-\text{PM}})^*(\Delta\phi_{\text{cd}-\text{PM}})}} \\ &= \sqrt{\frac{P_{\text{cd}}}{P_{\text{c}}}M_{\text{cd}}^2 \times \frac{1}{2}} \\ &= \sqrt{\frac{1}{2}\frac{T_{\text{sb}}2\bar{P}_{\text{sb}}P_{\text{cd}}}{P_{\text{c}}P_{\text{c}-\text{tot}}}}\end{aligned} \tag{9.9}$$

Combining Eqs. 9.6 and 9.9, the total rms-phase fluctuation due to the contrast defect and control sidebands is therefore

$$\begin{aligned}\tilde{\theta}_{\text{cd}} &= \sqrt{(\tilde{\theta}_{\text{c}-\text{AM}})^2 + (\tilde{\theta}_{\text{cd}-\text{PM}})^2} \\ &= \sqrt{\frac{T_{\text{sb}}2\bar{P}_{\text{sb}}P_{\text{cd}}}{P_{\text{c}}P_{\text{c}-\text{tot}}}}\end{aligned} \tag{9.10}$$

#### 9.1.2.2 Unbalanced Interferometer Control Sidebands

The effect of interferometer sidebands with unequal optical powers is shown in panels (d2) and (e2) of Fig. 9.3. The presence of a sideband power imbalance imposes a component of PM onto the carrier and AM onto the contrast defect. Calculated in a similar way as the contrast defect phase fluctuations, noting the switch between modulation type and carrier/contrast defect fields, the results are simply stated here. The modulation depth for the unbalanced sidebands is

$$M_{\text{unb}} = \frac{\delta E}{E_{\text{c}-\text{tot}}} = \sqrt{\frac{T_{\text{sb}}\delta P}{P_{\text{c}-\text{tot}}}} \tag{9.11}$$

The rms-phase fluctuation of the PM carrier field is

$$\begin{aligned}\tilde{\theta}_{\text{c}-\text{PM}} &= \sqrt{\overline{|\Delta\phi_{\text{c}-\text{PM}}|^2}} \\ &= \sqrt{\frac{P_{\text{cd}}}{P_{\text{c}}}M_{\text{unb}}^2 \times \frac{1}{2}} \\ &= \sqrt{\frac{1}{2}\frac{T_{\text{sb}}(\delta P)P_{\text{cd}}}{P_{\text{c}}P_{\text{c}-\text{tot}}}}\end{aligned} \tag{9.12}$$

The rms-phase fluctuation of the AM contrast defect field is

$$\begin{aligned}\tilde{\theta}_{\text{cd}-\text{AM}} &= \sqrt{\overline{|\Delta\phi_{\text{cd}-\text{AM}}|^2}} \\ &= \sqrt{\frac{P_{\text{cd}}}{P_{\text{c}}} M_{\text{unb}}^2 \times \frac{1}{2}} \\ &= \sqrt{\frac{1}{2} \frac{T_{\text{sb}}(\delta P) P_{\text{cd}}}{P_{\text{c}} P_{\text{c}-\text{tot}}}}\end{aligned} \tag{9.13}$$

The total rms-phase fluctuation due to interferometer control sidebands imbalance is

$$\begin{aligned}\tilde{\theta}_{\text{unb}} &= \sqrt{(\tilde{\theta}_{\text{c}-\text{PM}})^2 + (\tilde{\theta}_{\text{cd}-\text{AM}})^2} \\ &= \sqrt{\frac{T_{\text{sb}} \delta P}{P_{\text{c}} P_{\text{c}-\text{tot}}}}\end{aligned} \tag{9.14}$$

#### 9.1.2.3 Total Phase Noise Due to Interferometer Sidebands

The interferometer readout photodetector was used to measure the various optical powers. With a dark fringe offset set to zero, the contrast defect power was measured via length-scanning of the Output Mode Cleaner. The interferometer sidebands were measured by locking the Output Mode Cleaner to each sideband. The resulting values used in calculations are summarised in Table 9.2.

**Table 9.2** Measured values used in calculating phase noise contributions from the interferometer control sidebands

| Variable | Symbol | Value | Units | Calculation |
|---|---|---|---|---|
| OMC sideband transmission | $T_{\text{sb}}$ | 0.009 | – | Nominal |
| Contrast defect | $P_{\text{cd}}$ | 0.11 | mA | Measured |
| Carrier power | $P_{\text{c}}$ | 15 | mA | Measured |
| Total readout field | $P_{\text{c}-\text{tot}}$ | 15.11 | mA | $P_{\text{cd}} + P_{\text{c}}$ |
| Power in upper sideband | $P_{\text{u}}$ | 41 | mA | Measured |
| Power in lower sideband | $P_{\text{l}}$ | 46.3 | mA | Measured |
| Mean sideband power | $\bar{P}_{\text{sb}}$ | 43.5 | mA | $\|P_{\text{u}} + P_{\text{l}}\|/2$ |
| Difference in sideband power | $\delta P_{\text{sb}}$ | 5.3 | mA | $\|P_{\text{u}} - P_{\text{l}}\|$ |

The OMC readout photodiode was used to measure all power values

The rms-fluctuations due to the contrast defect and unbalanced sidebands was calculated to be $\tilde{\theta}_{\rm cd} = 20 \pm 0.5$ mrad and $\tilde{\theta}_{\rm unb} = 5 \pm 0.5$ mrad respectively. Combining Eqs. 9.10 and 9.14, the total phase fluctuations due to the interferometer sidebands is

$$\tilde{\theta}_{\rm sb} = \sqrt{\frac{T_{\rm sb} P_{\rm cd}}{P_{\rm c} P_{\rm c-tot}}(2\bar{P}_{\rm sb} + \delta P)} \tag{9.15}$$

and calculated to be $\tilde{\theta}_{\rm sb} = 21 \pm 1$ mrad.

### *9.1.3 Phase Noise of the Squeezed Interferometer*

The total phase noise of the interferometer with squeezing can be inferred from the squeezed quadrature of the nonlinear gain measurement made with the interferometer readout. Figure 9.4a shows the averaged squeezing level as measured by the interferometer in the 1.9–3.7 kHz band, as a function of the OPO nonlinear gain. The electronic gains of the squeezer control loops were adjusted to maintain unity-gain frequencies approximately constant. The interferometer remained in full lock throughout the this measurement, with no adjustments to the interferometer made. Along with 38 % detection efficiency, the total phase noise was inferred to

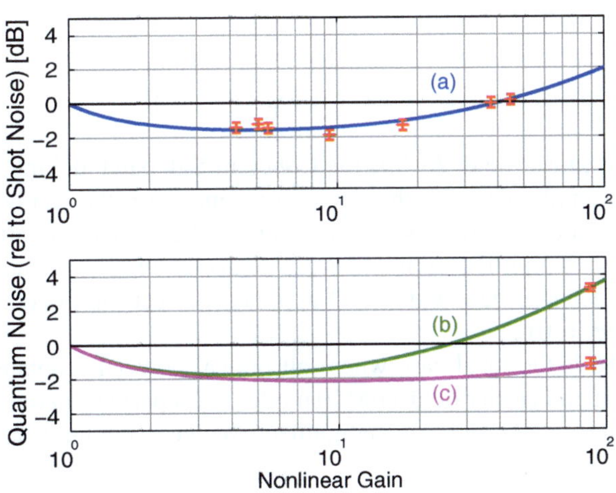


**Fig. 9.4** Squeezing as a function of OPO nonlinear, using interferometer readout. Trace *a* Squeezing quadrature measurement with varying OPO nonlinear gain operation, detection efficiency of 38 % and inferred phase noise of 85 ± 6 mrad. Trace *b* Pre-alignment adjusted squeezing quadrature measurement, with OPO nonlinear gain = 88, detection efficiency of 44 ± 2 %, and 108 ± 6 mrad phase noise. Trace *c* Post-alignment (optimised) squeezing quadrature measurement with OPO nonlinear gain = 88, detection efficiency of 44 ± 2 %, and 37 ± 6 mrad phase noise

be 85 ± 6 mrad. With a combined phase noise contribution from the squeezer and interferometer control sidebands of 30 ± 5 mrad, this measurement indicated there was another source of phase noise.

In a separate measurement run (as part of a different parameter measurement), the OPO was operated with a high nonlinear gain of 88. This measurement is shown in Fig. 9.4, Trace (b). A phase fluctuation level of 108 ± 6 mrad was inferred, substantially different to the measurement of Fig. 9.4a. However, while further optimising the interferometer operating state, it was noticed that the squeezing level changed with coherent locking demodulation phase and *interferometer alignment*. After a few iterations adjusting both interferometer alignment and coherent locking demodulation phase, the lowest squeezing level found corresponded to an inferred phase noise of 37 ± 6 mrad.

It was soon realised that *beam pointing at the AS port results in shifts of the coherent locking demodulation phase away from optimal squeezing measurement*, and this beam pointing contribution is explored next.

### 9.1.4 Beam Pointing and Phase Noise

As shown in the LIGO H1 Experiment schematic (Fig. 8.2), the Coherent Sideband locking photodetector for the LO loop was situated on the ISCT. The coherent locking photodetector measurement would result in a photocurrent given by

$$\begin{aligned} i_{CSL-IFO} &\propto |E_c + E^{CL}|^2 \\ &\propto (\alpha_c \cos(\omega_c t + \Phi) \times [\cos(\omega_c t + \Omega t) - A\cos(\omega_c t - \Omega t)])^2 \\ &\propto (\alpha_c\,[\cos(\Omega t - \Phi) - A\cos(\Omega t + \Phi)])^2 \end{aligned} \tag{9.16}$$

where $E_c$ is the carrier field, $E^{CL}$ is the Coherent Sideband locking field (Eq. 6.1), $\alpha_c$ is the carrier amplitude, $\omega_c$ is the carrier frequency, $A$ is the relative amplitude of the two coherent locking sidebands, and $\Omega$ is the coherent locking offset frequency. Equation 9.16 is a simplified version, having neglected the $2\omega_c$ terms filtered by the photodetector response bandwidth, and assumed perfect pump-coherent sideband field locking ($2\phi = 0$).

Equation 9.16 is the ideal case where all optical power is contained within the coherent locking and carrier field TEM00 modes, with perfect spatial overlap. However, higher order modes and/or misalignments between the fields modify Eq. 9.16 to become

$$i_{CSL-IFO} \propto \sum_{mn} \rho^{c}_{mn}\rho^{CL}_{mn}\,(\alpha_c[\cos(\Omega t-(\Phi_{00}+\phi_{mn}))-A\cos(\Omega t+(\Phi_{00}-\phi_{mn}))])^2 \tag{9.17}$$

The new variables are $\rho_{mn}$, the proportion of total optical power of a given field incident on the photodetector in the TEM m,n-th spatial mode, and $\phi_{mn}$, the difference between the relative phases of the TEM00 coherent-locking-carrier pair and m, n-th higher order coherent-locking-carrier pair. Demodulating the photocurrent (with in-phase and in-quadrature sinusoids $\cos(\Omega t+\phi_{dm})$ and $\sin(\Omega t+\phi_{dm})$), and electronic signal low-pass filtering (removing $2\Omega t$ terms), results in the in-phase $S_I$ and in-quadrature $S_Q$ Coherent Sideband locking error signals. These are of the form

$$S_I \propto \sum_{mn} \rho^c_{mn}\rho^{CL}_{mn}(\alpha_c[\cos(\phi_{dm}+\Phi_{00}-\phi_{mn}) - A\cos(\phi_{dm}+\Phi_{00}-\phi_{mn})])^2 \tag{9.18}$$

$$S_Q \propto \sum_{mn} \rho^c_{mn}\rho^{CL}_{mn}(\alpha_c[\sin(\phi_{dm}+\Phi_{00}-\phi_{mn}) - A\sin(\phi_{dm}+\Phi_{00}-\phi_{mn})])^2 \tag{9.19}$$

where $\phi_{dm}$ is the demodulating sinusoid phase. In the ideal case, $\rho_{00} = 1$ and $\phi_{00} = 0$, thus the ideal demodulation phase would be $\phi_{dm} + \Phi_{00} = 0$ or $\pi$. However, a change in beam alignments at the coherent locking photodetector changes the squeezing angle ($\theta_{sqz}$) away from optimum squeezing. For a set demodulation phase, the change in the squeezing angle from beam pointing is given by

$$\Delta\theta_{bp} \propto \frac{Q}{I} = -\frac{\sum_{mn} \rho^c_{mn}\rho^{CL}_{mn}(\alpha_c[\sin(\phi_{mn}) + A\sin(\phi_{mn})])^2}{\sum_{mn} \rho^c_{mn}\rho^{CL}_{mn}(\alpha_c[\cos(\phi_{mn}) - A\cos(\phi_{mn})])^2} \tag{9.20}$$

To illustrate the presence of this noise, a comparison between the noise spectra of the signal at the OMC and the Coherent locking error signal is shown in Fig. 9.5. The squeezing fluctuation spectrum at the interferometer output as measured by one

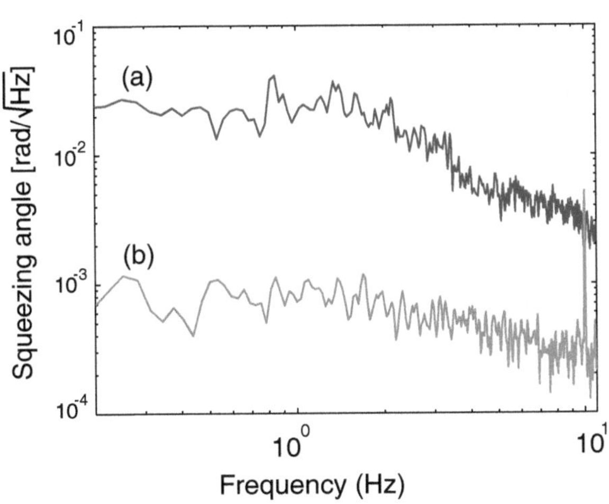


**Fig. 9.5** Direct measurement of squeezing ellipse noise. The squeezing fluctuation spectrum (Trace *a*) shows excess noise compared to the coherent locking error signal spectrum (Trace *b*), an indication of additional phase noise. Both signals are above the electronic noise floor of their respective photodetectors

**Table 9.3** Summary of squeezing ellipse phase noise results for the LIGO squeezed light injection experiment

| Phase noise source | Value (mrad) |
|---|---|
| Squeezer apparatus $\tilde{\theta}_{sqzr}$ | 22 ± 3 |
| Interferometer sidebands—contrast defect $\tilde{\theta}_{cd}$ | 20 ± 0.5 |
| Interferometer sidebands—imbalance $\tilde{\theta}_{unb}$ | 5 ± 0.5 |
| Interferometer sidebands—total $\tilde{\theta}_{sb}$ | 21 ± 1 |
| Squeezer + interferometer sideband noise | 30 ± 5 |
| Beam pointing (squeezing vs NL gain) | 79 ± 2 |
| Beam pointing (high NL gain) | 104 ± 2 |
| Beam pointing (high NL gain—optimised) | 22 ± 2 |
| **Total phase noise $\tilde{\theta}$ squeezer + sideband + beam pointing optimised** | 37 ± 6 |

of the reflection photodetectors of the Output Mode Cleaner, exceeds the squeezing fluctuations indicated by the Coherent locking error signal spectrum. This is evidence of the squeezing locking point fluctuating away from the optimum squeezing point.

### *9.1.5 Summary of Phase Noise Results*

A summary of phase noise results is shown in Table 9.3. The squeezing ellipse phase noise for the LIGO Squeezed Light Injection Experiment is $\tilde{\theta} = 37 \pm 6$ mrad. The contribution of beam pointing for the three measurements shown in Fig. 9.4 were calculated from the in-quadrature difference between the measured phase noise and other phase noise contributions (squeezer and interferometer sidebands). These beam pointing phase noise values should be regarded as upper limits in each measurement, but with values of $104 \pm 2$ mrad measured, this mechanism can be the dominating source of phase noise.

## 9.2 Interferometer Squeezing Detection Efficiency

The measurements made for assessing, then verifying, the squeezing detection efficiency of the interferometer injection are presented here. This completes the parameter characterisation of the squeezing injection experiment.

### *9.2.1 Squeezing/Antisqueezing as a Function of OPO Nonlinear Gain Using Interferometer Readout*

The squeezing detection efficiency can be inferred from the nonlinear gain measurement made with the interferometer readout. This was presented in Fig. 9.4, and a squeezing detection efficiency value of $\eta_{\text{det-IFO}} = 44 \pm 2\,\%$ was inferred. This value, measured with a *full operating interferometer*, is the nominal value to compare/verify.

### 9.2.2 Verification of the Interferometer Squeezing Detection Efficiency

To verify the squeezing detection efficiency inferred from the nonlinear gain measurement, two methods were used. These were:

i. Component measurement: Independent measurements of the optical losses of the components in the squeezing injection path and AS port path were made. These were subsequently combined together as a measure of the overall squeezing detection efficiency.
ii. Single-bounce measurement: A direct measurement of the squeezing injection path and AS port path using the interferometer single-bounce configuration. This configuration consists of deliberately misaligning the Y-arm test masses, X-arm ETM and the PRM. This removes the influence of the arm cavities and power-recycling cavity, making measurements of transmission/optical loss simpler. Two single-bounce configurations are shown in Fig. 9.6. The X-arm ITM acts as a

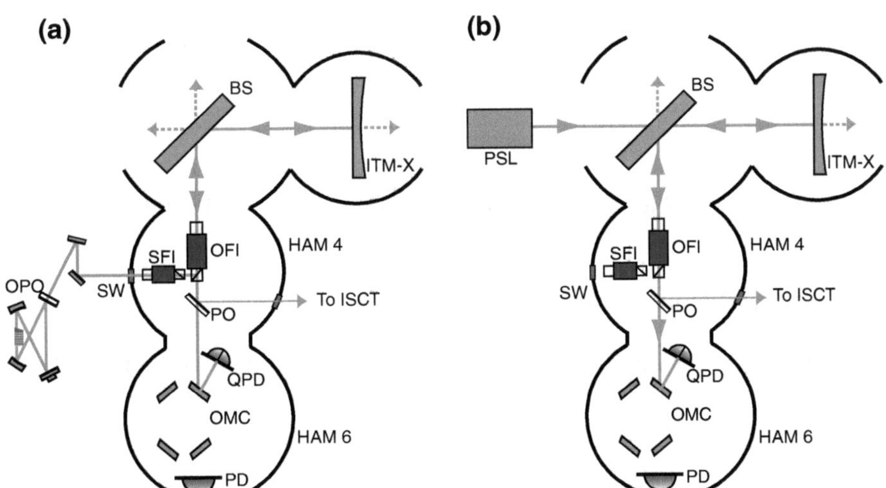


**Fig. 9.6** Illustration of the single-bounce mode configurations. **a** Using a bright beam from the squeezer and **b** Using a beam from the PSL. The OMC is shown off-resonance; the OMC can be locked on resonance to the injected beam to make efficiency measurements. *BS* interferometer beamsplitter; *ITM-X* input test mass of the X-arm; *OFI* output Faraday Isolator; *PO* pick-off mirror; *OMC* Output Mode Cleaner; *QPD* OMC reflection quadrant photodetector; *PD* interferometer readout photodetector, *OPO* Squeezer Optical Parametric Oscillator; *SW* Septum (injection) window; *ISCT* Interferometer Sensing and Control Table; *HAM* Horizontal Access Module vacuum chamber. Acronyms are the same as in Fig. 8.2

**Table 9.4** Additional interferometer parameters for single-bounce measurements

| Item | Symbol | Value |
|---|---|---|
| Power recycling mirror (PRM) transmission | $T_{PRM}$ | 0.027 |
| Input test mass (ITM) reflectivity | $R_{ITM}$ | 0.97 |
| Beamsplitter (BS) Transmission | $T_{BS}$ | 0.25 |
| Carrier AS port power fraction (from the PSL) | $Z_{SB}$ | 0.965 |

single reflector, thus optical power measurements of the AS port can be made by factoring the known reflectivity of ITM-X ($R_{ITM}$) and the 25 % transmission of double-passing the Beamsplitter ($T_{BS}$). Additional nominal parameters for single-bounce measurement are given in Table 9.4.

The results of squeezing efficiency in HAM4 and HAM6 chambers are presented next. These results were measured using both component and single-bounce methods.

## 9.2.3 Injection Hardware Optical Losses in HAM4

### 9.2.3.1 Component Measurement

Before installation, the transmission loss of the Squeezing Faraday Isolator (SFI) and the Output Faraday Isolator (OFI) in both propagation directions were measured in separate benchtop experiments. The results of these measurements are presented in Table 9.5, along with quoted values for the injection septum window and pick off mirror. The transmission efficiency for the HAM4 hardware using the component measurement method $\eta_{HAM4-com}$ is thus:

$$\eta_{HAM4-com} = T_{SW} \times T_{SFI} \times T_{OFIrev} \times T_{OFI} \times T_{PO} = 0.84 \pm 0.03 \qquad (9.21)$$

### 9.2.3.2 Single-Bounce Measurement

In the interferometer single-bounce configuration shown in Fig. 9.6a, a bright beam of 6.5 mW was injected from the squeezer OPO. This resulted in 1.31 mW being

**Table 9.5** Parameters of the different HAM4 hardware components

| Item | Symbol | Power transmission |
|---|---|---|
| Injection (septum) window | $T_{SW}$ | $0.998 \pm 0.002$ |
| Squeezing injection Faraday isolator | $T_{SFI}$ | $0.94 \pm 0.02$ |
| Output Faraday isolator (reverse) | $T_{OFIrev}$ | $0.966 \pm 0.01$ |
| Output Faraday isolator (forward) | $T_{OFI}$ | $0.94 \pm 0.02$ |
| HAM4 pickoff mirror to ISCT | $T_{PO}$ | $0.988 \pm 0.01$ |

registered on the OMC reflection photodiode. Factoring out the interferometer beam-splitter and reflectivity of the ITM, the transmission efficiency measured in single-bounce mode $\eta_{\mathrm{HAM4-sbm}}$ is

$$\begin{aligned} P_{\mathrm{QPD}} &= 1.31 \pm 0.05\ \mathrm{mW} \\ P_{\mathrm{in}} &= 6.5 \times T_{\mathrm{BS}} \times R_{\mathrm{ITM}} = 1.58 \pm 0.05\ \mathrm{mW} \\ \eta_{\mathrm{HAM4-sbm}} &= P_{\mathrm{QPD}}/P_{\mathrm{in}} = 0.83 \pm 0.04 \end{aligned} \tag{9.22}$$

The single-bounce value is consistent with the HAM4 component measurement value (Eq. 9.21). We take the $\eta_{\mathrm{HAM4-sbm}}$ as the transmission efficiency for the HAM4 optical components.

### 9.2.4 Optical Losses in HAM6—Output Mode Cleaner Mode-Matching Losses and Transmission Throughput

The measurements of mode-matching and transmission throughput of the OMC were made using single-bounce configurations.

A measurement of the mode-matching of the squeezed beam spatial mode to the OMC cavity mode was made using a bright field from the squeezer in single-bounce. Looking at the OMC reflection photodetector (QPD) while scanning the OMC optical path length, the measured optical power measured when the OMC went through resonance dropped to 26 % of the off-resonance optical power level [2]. This infers that the mode-matching of the squeezed beam spatial mode to the OMC cavity mode is $\eta_{\mathrm{MM}} = 74 \pm 2\,\%$.

The OMC mode-matching efficiency for the squeezed beam was anticipated to be greater than 90 %. The process to mode-match the squeezed beam to the OMC cavity mode was made with optical lenses placed on the squeezer optical table. The path length between the mode-matching optics and the OMC (~60 m) made optimising the optics' placement difficult. The slow scanning of the OMC cavity optical path length also hindered the speed of the mode-matching process, reducing the number of optimisation runs that could be undertaken in the Squeezing Experiment time-window. Recommendations to improve the mode-matching process will be made in Sect. 9.7. The mode-matching is also affected by the thermal heating of the full interferometer. The thermal heating effect on modematching will be discussed in Sect. 9.2.6.

The throughput of the OMC was measured in single-bounce mode with a 2 W input beam from the PSL. This single-bounce configuration is shown in Fig. 9.6b. This resulted in 8.7 mW of light incident on HAM6. With the OMC locked on resonance, the interferometer readout photodiode measured 6.43 mW of power. Including the nominal mode-matching coefficient $T_{\mathrm{nom-MM}} = 0.93$ [3], and the Carrier AS port power fraction $Z_{\mathrm{SB}} = 0.965$ [3], the throughput of the OMC $\eta_{\mathrm{OMC}}$, including the readout photodetector efficiency is

$$\eta_{\text{OMC}} = \frac{6.43}{8.7 \times T_{\text{nom-MM}} \times Z_{\text{SB}}} = 82.5 \pm 0.4\,\% \tag{9.23}$$

For squeezing enhancement, the ideal throughput of the OMC should be near 100 %, that is, the OMC should be impedance matched for the squeezing. It was expected that the throughput would be greater than 90 % after the swap-installation of the L1 OMC (Sect. 8.2.2). The reasoning behind the loss in throughput of the Output Mode Cleaners remains an open area of investigation (see Appendix B for initial testing of the H1 OMC).

## *9.2.5 Results of Squeezing Detection Efficiency Verification Tests*

### 9.2.5.1 Combined Component Measurement

Combining the transmission, modematching and throughput efficiencies of the optical components in HAM4 and HAM6, the calculated squeezing detection efficiency for the interferometer, $\eta_{\text{det-IFO-com}}$, is therefore

$$\eta_{\text{det-IFO-com}} = \eta_{\text{HAM4-sbm}} \times \eta_{\text{MM}} \times \eta_{\text{OMC}} = 0.51 \pm 0.04 \tag{9.24}$$

The squeezing detection efficiency measured by component method is 51 ± 4 %.

### 9.2.5.2 Single Bounce Measurement

A single-bounce measurement locking the OMC to a bright field from the squeezer was made. With an injected power of 6.5, 0.83 mW was measured on the readout photodetector. Therefore the squeezing detection efficiency as measured in single-bounce mode, $\eta_{\text{det-IFO-sbm}}$, is

$$\eta_{\text{det-IFO-sbm}} = \frac{0.83}{6.5 \times T_{\text{BS}} \times R_{\text{ITM}}} = 0.52 \pm 0.04 \tag{9.25}$$

The single-bounce value of the squeezing detection efficiency 52 ± 4 %. This agrees well with the combined component value $\eta_{\text{det-IFO-com}}$, thus the verification results are consistent.

## *9.2.6 Comparison to the OPO Nonlinear Gain Result*

The values of the squeezing detection efficiency from the verification tests are higher than the OPO nonlinear gain measurement $\eta_{\text{det-IFO}} = 44 \pm 2\,\%$. The difference in squeezing detection efficiency between measurement types is credited to the thermal effects of the optical circulating power with the full interferometer. Effects such as thermal deformation of the optics, for example, changes in the ITM radius-of-curvature, cause changes to the spatial mode of the squeezed beam being reflected. This effect impacts the mode-matching efficiency of the squeezed beam to the OMC, $\eta_{MM}$. Due to only using low optical power, the component and single-bounce measurements do not experience the thermal effects.

As the OPO nonlinear gain measurement was made with a fully operating interferometer (that is, with the full thermal effects of circulating optical power), $\eta_{det-IFO} = 44 \pm 2\,\%$ is taken to be the squeezing detection efficiency achieved for the LIGO Squeezed Light Experiment. The verification tests still proved very useful in identifying areas of improvement needed to achieve higher detection efficiency overall.

## 9.3 Squeezing-Enhanced Interferometer Sensitivity

The LIGO squeezed light source was operated with 50 mW of pump power. The generated squeezing was injected into the open port of the Output Faraday isolator, acting as the optical circulator. The coherent locking loops were active, controlling the squeezing ellipse phase to the interferometer readout quadrature. The interferometer was operated with 20 W of input laser power from the PSL. This was the same amount of input power used during the joint scientific data run [4]. The strain-sensitivity curve of the interferometer with no squeezing is named the 'Reference' curve.

Figure 9.7 shows the measured Reference trace, and a trace with injected squeezing measured shortly after. Squeezing-enhancement of the shot noise dominated region of up to $2.15 \pm 0.05$ dB is observed, and has improved the overall sensitivity trace above 150 Hz. This represents the first measurement of squeezing in a 4 km LIGO gravitational-wave detector. Figure 9.8 shows the data of Fig. 9.7, focussed into the region where Enhanced LIGO has the best strain-sensitivity to gravitational waves. There is squeezing enhancement of the strain-sensitivity in this region.

Figure 9.9 shows the dB-improvement relative to the Reference curve. Above 1 kHz, applied squeezing is reducing the effect of quantum shot noise in detector readout. The loss of squeezing magnitude at lower frequencies is indicative of the thermal noise and seismic noise becoming dominant noise sources over shot noise. However, squeezing improvement is still observable down to 150 Hz.

## 9.4 Surpassing the Joint-Scientific Data Run Sensitivity Above 250 Hz

The Reference and Squeezing curves of Fig. 9.7 have undergone a measurement-calibration process. This process calibrates strain sensitivity data from a gravitational-wave detector with applied calibration frequency signals. The joint scientific run data

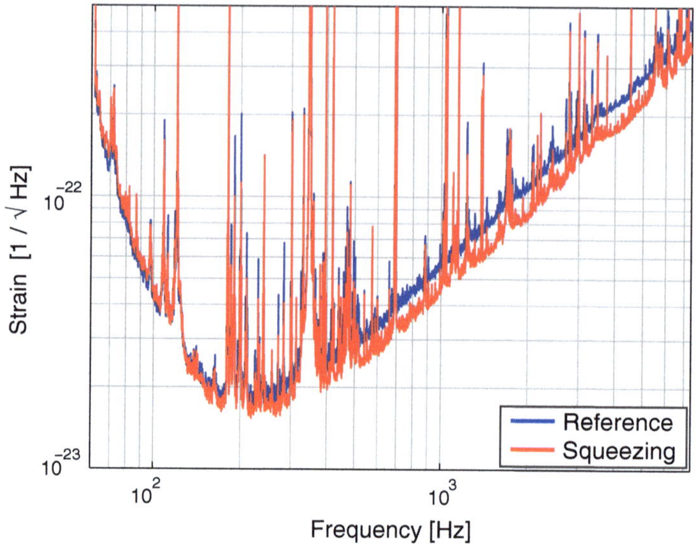


**Fig. 9.7** Enhanced LIGO with squeezing. The Reference trace shows the sensitivity of the H1 readout without squeezing being injected, while the Squeezing trace shows the interferometer readout with squeezing injected

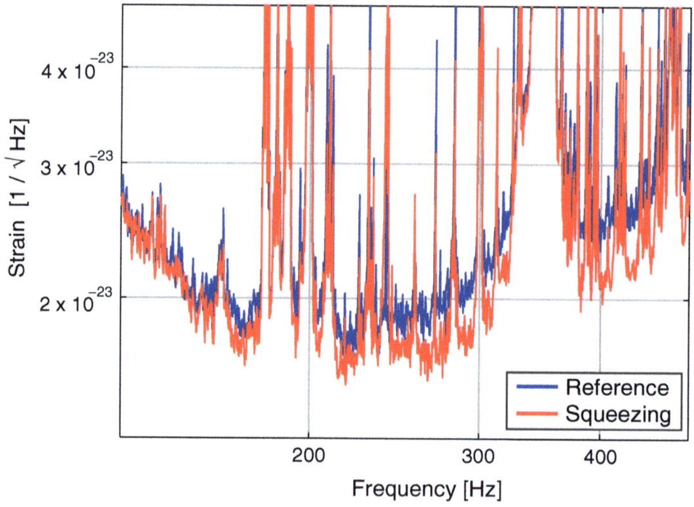


**Fig. 9.8** Squeezing enhancement in LIGO's most sensitive frequency band, between 150 and 300 Hz

[5] is calibrated with the same process. We can therefore compare the LIGO Squeezed Light Injection sensitivity curves with the Hanford Interferometer data from the joint scientific run.

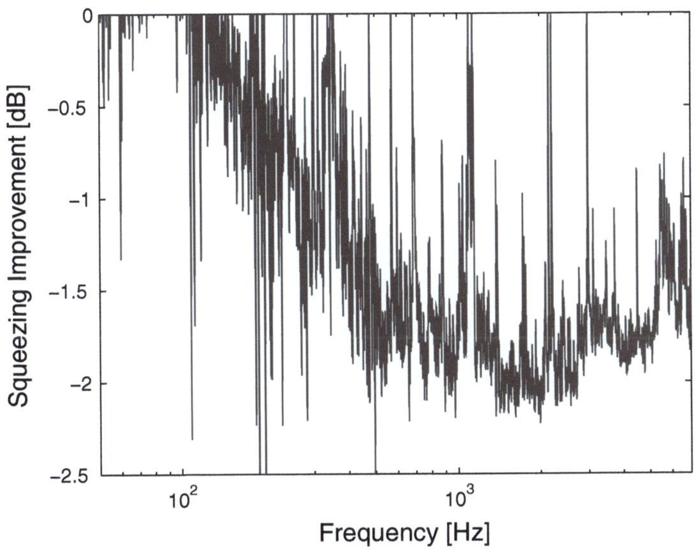


**Fig. 9.9** Squeezing improvement measured, relative to the reference spectrum. A maximum improvement of $2.15 \pm 0.05$ is observed around the 1.5 kHz region, and squeezing improvement is still observable at 150 Hz

Figure 9.10 shows the Squeezing Experiment strain-sensitivity traces with the eLIGO H1 sensitivity measured during the joint scientific run. The squeezing trace broadly surpasses the scientific-run sensitivity above 250 Hz, therefore represents the best sensitivity in a gravitational-wave detector above 250 Hz yet achieved.

Comparing the LIGO Squeezed Light Injection Experiment data with the joint scientific run data, there are differences in features. Firstly, below 150 Hz, the Reference and Squeezing traces are higher than the joint science run trace. The higher noise level is due to a difference in the seismic environment, namely that the Squeezing Experiment was conducted in a seismically-noisier period than when the joint-scientific data was measured. Secondly, there are more peaks in the Squeezing Experiment data. This is credited to higher local environment noise during the Squeezing Experiment. The local environment of the squeezer was subject to the activities of the installation work for Advanced LIGO. Great effort by the LIGO Hanford Observatory staff and LIGO Squeezed Light Injection Team were made to make the environment as quiet as possible during the Squeezed Light Injection Experiment, however, halting of all Advanced LIGO installation activities was not possible.

## 9.5 Squeezing and Interferometer Input Power

The squeezing improvement can be assessed in terms of equivalent interferometer input power. This is shown in Fig. 9.11. The Reference and Squeezing traces are as before. The Scaled trace is the Reference data (at 20 W input power) scaled to

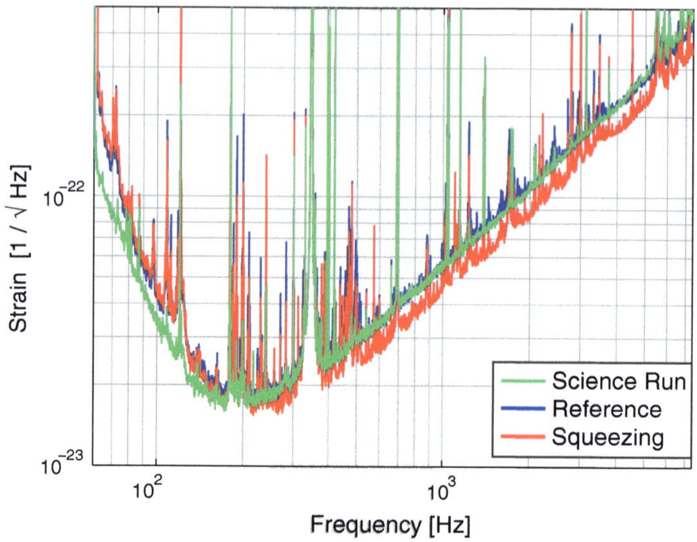


**Fig. 9.10** Squeezing improvement compared to the joint scientific run sensitivity. The difference at the lower frequencies is due to the differing seismic environments. The Reference trace broadly matches the scientific run trace above 250 Hz. Therefore the Squeezing trace broadly surpasses the joint scientific run sensitivity above 250 Hz

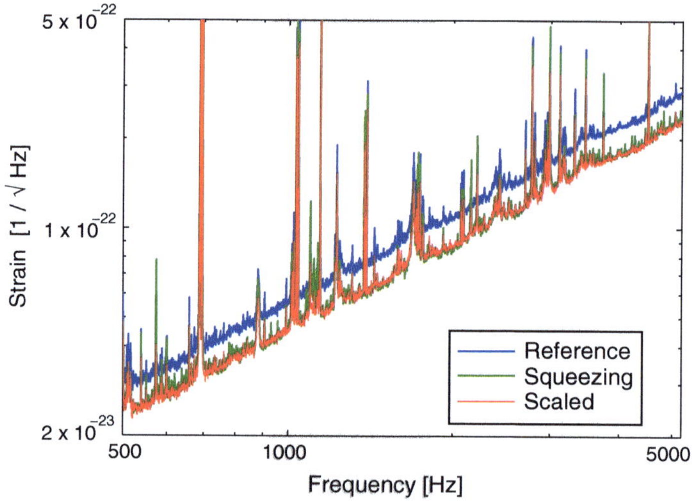


**Fig. 9.11** Squeezing enhancement and equivalence to higher input power. The Reference and Squeezing traces are as before. The *Scaled trace* is the Reference data (at 20 W input power) scaled to be equivalent shot noise for 31 W input interferometer power. This shows the equivalence of lower input power (20 W) with squeezing and higher input (31 W) with no squeezing

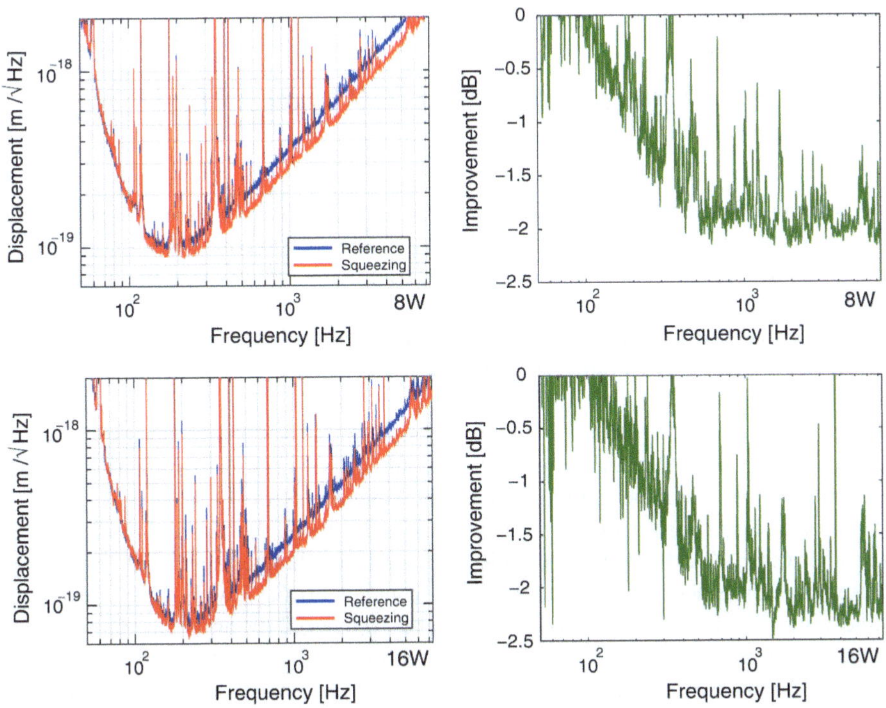


**Fig. 9.12** Spectra with and without squeezing, with 8 W (*top row*) and 16 W (*bottom row*) input power into the interferometer

be equivalent shot noise for 31 W input interferometer power. The Squeezing and Scaled traces are now superimposed. This shows the equivalence of lower input power (20 W) with squeezing, and higher input power (31 W) with no squeezing. The result supports the claim that injected squeezed states can be used to regain similar quantum noise performance with lower interferometer-operating laser powers if needed (Sect. 9.1).

Squeezing injection was also applied to interferometer input powers below 20 W. Figure 9.12 shows the interferometer spectra and noise reduction due to squeezing when the interferometer input power was at 8 and 16 W respectively. These spectra have not been callibrated, thus cannot be directly compared to equivalent earlier data of the joint scientific run. With different input optical powers, squeezing-enhancement is still observed above 150 Hz.

## 9.6 Inferred Squeezing Improvement

### *9.6.1 Inferred Squeezing from the LIGO Squeezed Light Source*

We infer the squeezing level exiting the LIGO H1 OPO cavity, using the squeezing and antisqueezing data measured on the diagnostic homodyne detector, shown in Fig. 8.6. This is the inferred amount of squeezing being injected into the interferometer.

The propagation efficiency between the OPO and the diagnostic homodyne was $\eta_{prop} = 98 \pm 1\,\%$. The homodyne efficiency was measured to be $\eta_h = 95 \pm 1.0\,\%$, and photodetector efficiency was $91 \pm 2\,\%$. Combining these efficiencies, the detection efficiency is $\eta_{det} = 85 \pm 4\,\%$. After accounting for the OPO escape efficiency of $\eta_{esc-H1} = 95.9\,\%$, the detection efficiency agrees with a squeezing/antisqueezing as a function of nonlinear gain measurement, shown in Sect. 9.1.1.

Firstly, correcting for the homodyne dark noise level, the squeezing magnitude becomes $6.3 \pm 0.2$ dB, while the antisqueezing level changes negligibly. Secondly, correcting for squeezing ellipse phase noise using Eq. 6.17 with $22 \pm 3$ mrad of squeezing ellipse phase noise (Sect. 9.1.1) and $10.6 \pm 0.2$ dB antisqueezing, gives a squeezing level of $6.4 \pm 0.3$ dB. Finally, correcting for the detection efficiency using Eq. 6.6, the inferred squeezing level exiting the OPO cavity is $10.3 \pm 0.4$ dB.

### *9.6.2 Inferred Level of Squeezing-Sensitivity Enhancement*

We now infer the squeezing improvement expected using the Experiment parameters. Beginning with the inferred value of $10.3 \pm 0.4$ dB squeezing leaving the OPO, this is corrected for the squeezing detection efficiency of $\eta_{\mathrm{det-IFO}} = 44 \pm 2\,\%$, using Eq. 6.6 on both quadratures. This is then corrected for the squeezing ellipse phase noise of $37 \pm 6$ mrad (Sect. 9.1.5), using Eq. 6.17.

The expected squeezing-improvement to be observed is calculated to be $2.15 \pm 0.37$ dB. With a measured squeezing improvement of up to $2.15 \pm 0.05$ dB shown in Fig. 9.9, the interferometer readout measurement and predicted sensitivity improvement are in good agreement.

## 9.7 Improvements for applying Squeezing in Future Gravitational-Wave Detectors

Reducing squeezing ellipse phase noise and increasing squeezing detection efficiency will lead to greater squeezing-enhancement of future gravitational-wave detectors. This section presents possible methods to improve these squeezing-injection parameters.

### 9.7.1 Reducing Sources of Squeezing Ellipse Phase Noise

We divide the solutions for reducing the squeezing ellipse phase noise into the sources of phase noise.

- **From the squeezer apparatus**—Phase noise on the squeezer table is dominated by path length fluctuations of optical beams. Mechanisms for these fluctuations include acoustic noise and air currents. Possible methods to improve the phase noise include enclosing critical beam paths, covering the squeezer apparatus with airtight enclosures, and moving the squeezer OPO into the interferometer vacuum envelope.
- **From the interferometer control sidebands**—Recalling Eq. 9.15 for the interferometer sidebands contribution to phase noise

$$\tilde{\theta}_{\text{sb}} = \sqrt{\frac{T_{\text{sb}} P_{\text{cd}}}{P_{\text{c}} P_{\text{c-tot}}} (2\bar{P}_{\text{sb}} + \delta P)}$$

  it is observed that should the ratio of the carrier amplitude to contrast defect be increased (i.e. $P_{\text{cd}} \ll P_{\text{c}}$), the phase noise will be reduced. Minimising the proportion of contrast defect field power in the interferometer readout will minimise the interferometer sidebands contribution to phase noise.
- **From beam pointing fluctuations**—An active alignment scheme should be installed to minimise beam pointing fluctuations. Readout of the beam alignment can be achieved using quadrant photodiodes measuring the relative beam pointing between the interferometer carrier field and squeezing coherent control sidebands. Actuation on the squeezed beam can be fed back to the two PZT steering mirrors on the squeezer table (Sect. 8.3).

### 9.7.2 Improving Squeezing Detection Efficiency

- **Component optical loss**—Improving optical loss in components of the squeezing injection path will be a considerable research/development undertaking. Increasing the transmission efficiency of critical components such as Faraday isolators, and increasing the Output Mode Cleaner throughput will be needed in order to access higher levels of applied squeezing.
- **Spatial Modematching to the OMC**—The first improvement is to increase the OMC optical path length scan rate. This then allows many more iterations to assess the modematching of the squeezed beam to the OMC cavity mode. This improvement is part of Advanced LIGO OMC design specifications [6].
  The second improvement is to replace the fixed modematching lens system with a tunable modematching system as part of the squeezing injection path. Tunable modematching have been demonstrated using heated-deformable mirrors [7]. This

is also being developed for modematching the interferometer output beam to the Output Mode Cleaner [8]. A tunable modematching system potentially offers the ability to precisely modematch the squeezed beam to the OMC, and the ability to in-situ alter the modematching to compensate for changes from thermal heating of the full interferometer.

## 9.8 Summary

The parameters of the LIGO Squeezed Light Injection Experiment that impact the observed squeezing-enhancement were identified and quantified. The results of squeezing ellipse phase noise, originating from contributions of the squeezer apparatus, interferometer control sidebands, and inferred beam pointing phase noise were reported. The squeezing detection efficiency was measured, then verified using two characterisation methods.

This chapter then reported the squeezing-enhancement of the Enhanced LIGO detector, with an measured improvement of up to $2.15 \pm 0.05$ dB down to 150 Hz. This was the first observation of squeezing-enhancement of a gravitational-wave detector surpassing limiting quantum shot noise. The presence of a squeezed-light source added no additional noise to low frequency audio-detection band sensitivity. Above 250 Hz, this represents the best sensitivity in a gravitational-wave detector yet achieved. The capability of squeezing to recover detector-sensitivity using lower interferometer input power was also demonstrated.

The expected squeezing enhancement was inferred from the squeezing level generated by the OPO, correcting for the squeezing detection efficiency and squeezing ellipse phase noise. The measured squeezing enhancement and the expected squeezing enhancement were found to be in good agreement. Recommendations for improving the application of squeezed states in future gravitational-wave detectors concluded the Chapter.

## References

1. K. Somiya, Y. Chen, S. Kawamura, N. Mio, Frequency noise and intensity noise of next-generation gravitational-wave detectors with RF/DC readout schemes. Phys. Rev. D **73**, 12 (2005)
2. S. Dwyer, Mode scan with our beam. Hanford iLog (2011)
3. L. Barsotti, Summary of measured losses. Hanford iLog (2011)
4. G.M. Harry (for the LIGO Scientific Collaboration), Advanced LIGO: the next generation of gravitational wave detectors. Class. Quantum Gravity **27**, 084006 (2010)
5. The LIGO Scientific Collaboration and The Virgo Collaboration (J Abadie et al.), Sensitivity achieved by the LIGO and virgo gravitational wave detectors during LIGO's sixth and virgo's second and third science runs (2012), arXiv:1203.2674v2

6. K. Arai, S. Barnum, P. Fritschel, J. Lewis, S. Waldman, Output mode cleaner design, LIGO-T1000276-v5 (2013)
7. M.A. Arain, W.Z. Korth, L.F. Williams, R.M. Martin, G. Mueller, D.B. Tanner, D.H. Reitze, Adaptive control of modal properties of optical beams using photothermal effects. Opt. Express **18**(3), 2767–2781 (2010)
8. N. Smith-Lefebvre, Techniques for improving the readout sensitivity of gravitational wave antennae. Ph.D. thesis, Department of Physics, Massachusetts Institute of Technology, Cambridge, 2012

# Chapter 10
# Backscattered-Light Impact in a Squeezing-Enhanced Gravitational-Wave Detector

**Abstract** This chapter reports the investigations that concluded, from a backscattered-light perspective, the presence of a squeezer is not a fundamental detriment to the strain-sensitivity of a gravitational-wave detector. The impact of backscattered light in gravitational-wave detector readout is first introduced in Sects. 10.1 and 10.2. After overviews of the LIGO Squeezed Light Injection Experiment environment conditions (Sect. 10.3) and the evidence of the backscattered-light mechanism (Sect. 10.4), the results of two backscatter experiments are presented. These are the large displacement (Sect. 10.5) and small displacement measurements (Sect. 10.7). The bi-directional scatter distribution function of the H1 Optical Parametric Oscillator is estimated from the large displacement measurement result (Sect. 10.6), and the consistency between the two experiments is discussed in Sect. 10.8. Recommendations to further reduce the backscattered light impact of an installed squeezer in future gravitational-wave detectors conclude the chapter (Sect. 10.9).

## 10.1 Backscattered Light from the Interferometer as Port

The impact of backscattered light is of serious concern in the potentially source-rich lower frequencies of the gravitational-wave detection band, up to 300 Hz (Sect. 2.1). Backscattered light can create spurious signals that mask gravitational-wave signals [1–3].

Figure 10.1 shows a simplified schematic of the interferometer anti-symmetric (AS) port with the readout and installed squeezer. Light from the AS port field can scatter from the interferometer readout path up into the squeezing injection path towards the squeezer. This light can then be back-reflected by the squeezer to co-propagate with the injected squeezing, thus reaching the readout photodetector. For the purpose of this chapter, this is what is referred to as backscattered light.[1] This definition of backscattered light is similar to that in Chap. 7, with the measurement Local Oscillator being the interferometer AS port field.

[1] Scattered light from other parts of the interferometer are also of concern, see [5–7]. However, the focus is the AS port—the source of scattered light that can be backscattered by an installed squeezer.



S.S.Y. Chua, *Quantum Enhancement of a 4 km Laser Interferometer Gravitational-Wave Detector*, Springer Theses,
DOI 10.1007/978-3-319-17686-4_10

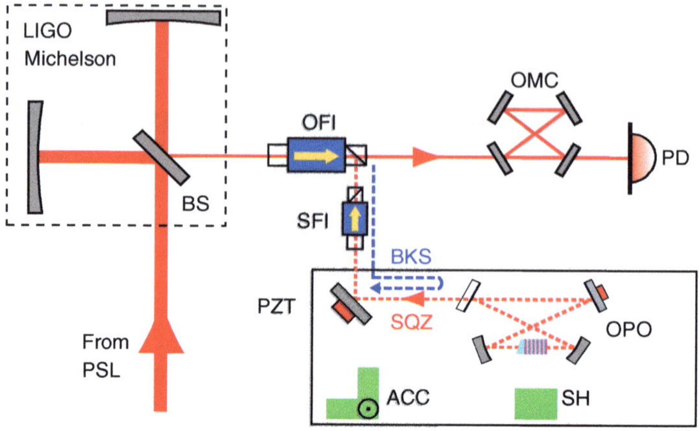


**Fig. 10.1** Simplified schematic showing the squeezer, the interferometer and the backscattered-light path. The backscattered light (*BKS*) optical path of interest is shown. *PSL* Pre-stabilised laser; *BS* interferometer beamsplitter; *OFI* output Faraday isolator; *SFI* squeezing-injection Faraday isolator; *OMC* output mode cleaner, *PD* readout photodetector. Equipment used in the tests: *PZT* injection path piezo-electric transducer; *SH* shaker unit, *ACC* accelerometers. Reprinted with permission from [4], copyright (2014) by IOP Publishing

## 10.2 Theoretical Derivations and Experiments Overview

The derivation of the theoretical backscattered-light equation is presented here, along with a brief overview of the backscatter experiments undertaken as part of the LIGO Squeezed Light Injection Experiment.

### 10.2.1 Backscattered-Light Theory

We begin by denoting the incident power on the interferometer readout photodetector due to the carrier as $P_c = |E_c|^2$, the backscatter as $P_s = |E_s|^2$ ($\ll P_c$), and their relative phase as $\phi_s$. This is illustrated in Fig. 10.2a, along with the equivalent phasor

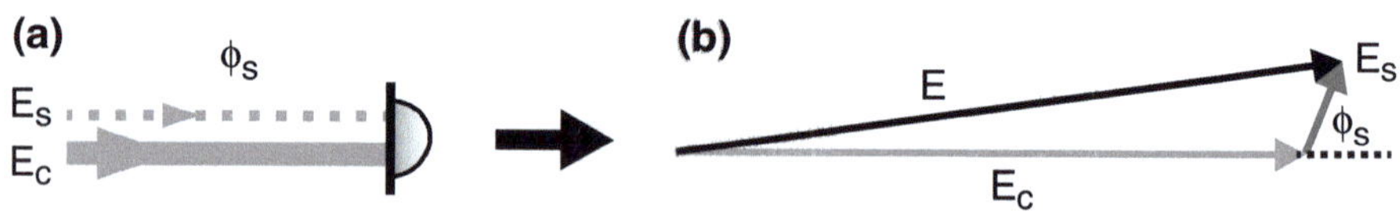


**Fig. 10.2** **a** Simple illustration of beams incident on the readout photodetector. **b** Phasor diagram showing the relationships of the various fields. Reprinted with permission from [4], copyright (2013) by IOP Publishing

diagram of the optical fields in Fig. 10.2b. The carrier power can be expressed as steady-state and fluctuating components $P_c = \bar{P}_c + \delta P_c$, where $\delta P_c \ll \bar{P}_c$.

The total power detected $P$ is given by

$$\begin{aligned} P = E^2 &= E_c^2 + E_s^2 - 2E_c E_s \cos(\pi - \phi_s) \\ &= P_c + P_s + 2\sqrt{P_c P_s}\cos(\phi_s) \\ &\approx \bar{P}_c + 2\sqrt{\bar{P}_c P_s}\cos(\phi_s) \\ &= \bar{P}_c + \mathrm{dS} \end{aligned} \tag{10.1}$$

where

$$\mathrm{dS} = 2\sqrt{\bar{P}_c P_s}\cos(\phi_s) \tag{10.2}$$

is the *backscatter noise contribution in the interferometer readout*. The relative phase $\phi_s$ is assumed to accrue entirely from the total beam-path displacement due to the scattering object $X$, via the relation

$$\phi_s = 2kX = \frac{4\pi}{\lambda}X \tag{10.3}$$

with the wavenumber $k = 2\pi/\lambda$. The total beam-path displacement can be written as two terms that describe the contributions from large beam-path displacements, $X_s$, and small beam-path displacements, $\delta x_s (\ll X_s)$, namely $X = X_s + \delta x_s$. Equation 10.2 can now be expressed in terms of *relative intensity noise* (RIN), namely

$$\begin{aligned} \mathrm{RIN}_s = \frac{\mathrm{dS}}{\bar{P}_c} &= 2\sqrt{\frac{P_s}{\bar{P}_c}}\cos(\phi_s) \\ &= 2\sqrt{\frac{P_s}{\bar{P}_c}}\cos(2k(X_s + \delta x_s)) \\ &\approx 2\sqrt{\frac{P_s}{\bar{P}_c}}\left[\cos(2kX_s)\cos(2k\delta x_s) + \sin(2kX_s)\sin(2k\delta x_s)\right] \\ &\approx \underbrace{2\sqrt{\frac{P_s}{\bar{P}_c}}\cos\left(\frac{4\pi X_s}{\lambda}\right)}_{(i)} + \underbrace{\sqrt{\frac{2P_s}{\bar{P}_c}}\left(\frac{4\pi\delta x_s}{\lambda}\right)}_{(ii)} \end{aligned} \tag{10.4}$$

Equation 10.4 is reached by approximating $\cos(2k\delta x_s) \approx 1$, $\sin(2kX_s) \approx 1/\sqrt{2}$ over many cycles. This shows the backscatter noise contribution in the interferometer readout is separable into large displacement and small displacement terms, and that is dependent on the amount of (DC) optical power of the backscattered beam ($P_s$), and the motion of the scattering object ($X$).

### 10.2.2 Scattered Light and OPO Nonlinear Gain

Any scattered light back-reflected *within the OPO* will experience the OPO nonlinear parametric gain, in the same way as a forward-propagating seed field. This is an additional process that needs to be accounted for that impacts the scattered light power $P_s$.

Amplification or deamplification will be determined by the relative phase between the pump and the scattered fields. As in Sect. 8.6, the OPO nonlinear gain effect on seed field optical power can be accounted for by

$$P_{\mathrm{out}}^{0} = P_{\mathrm{out}}^{\pm} \times \left(\frac{1 \pm x}{1 - x^2}\right)^{-2} \tag{10.5}$$

with $P_{\mathrm{out}}^{0}$ the optical power with no nonlinear gain, $P_{\mathrm{out}}^{\pm}$ the optical power with amplification/deamplification, and the normalised pump parameter $x$. The normalised pump parameter is related to the OPO classical nonlinear gain $G$ by

$$x = 1 - \frac{1}{\sqrt{G}} \tag{10.6}$$

## 10.3 Experiment Conditions

### 10.3.1 Installed Hardware for Backscatter Noise Mitigation

To mitigate the potential backscatter impact, the bow-tie OPO cavity with its intrinsic backscatter isolation was combined with a Faraday isolator (SFI) placed in the squeezing-injection path to provide additional isolation.

The OPO nonlinear crystal was from the same manufacturing run as the nonlinear crystal contained within the DB-OPO at ANU. Further, the LIGO OPO was constructed with cavity mirrors of higher-quality surface polishing and optical coatings (Sect. 8.4). Therefore, it is expected that the intrinsic isolation of the LIGO OPO is greater to that measured for the DB-OPO. The bi-directional scatter distribution function (BSDF) of the LIGO OPO will be inferred from results presented in this Chapter, and reported in the Sect. 10.6.

The squeezing-injection Faraday isolator had a backscatter isolation of 37 dB, when initially tested in a bench-top experiment. However, when installed into the squeezing-injection path, a separate measurement of returning optical power provided evidence that the isolation decreased to 23 dB [8]. The cause of this decrease is unknown.

Lastly, the Output Faraday isolator (OFI) had a measured attenuation of reflected optical power into the squeezing injection path of 23 dB [9].

### 10.3.2 Seismic and Acoustic Environment of the Squeezer Table

The seismic and acoustic hardware of the squeezer table are presented in Fig. 10.3. The squeezer was elevated on optical table tripod legs, so as to obtain the necessary height (relative to the interferometer beam path) for injecting squeezed light into the interferometer. For additional seismic damping, the tripod legs were grouted to the floor. A measurement of the background displacement motion spectra is presented in Fig. 10.4.

The squeezer table was covered with an enclosure. The enclosure panels were made from lightweight plastic board with a thin metal outer layer. A sample of the enclosure material is also shown in Fig. 10.3. The panels, approximately 5 mm

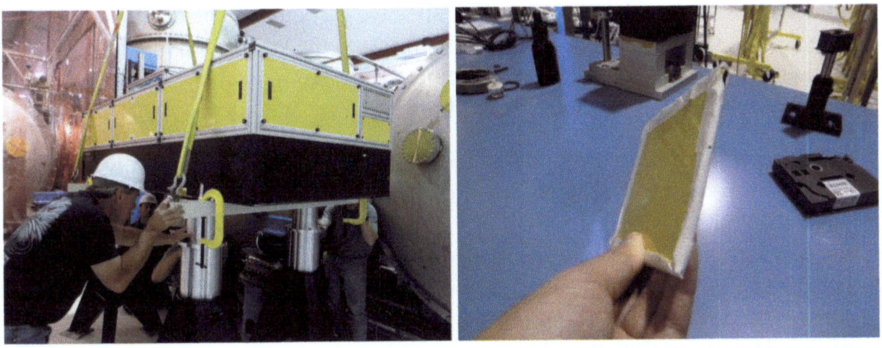

**Fig. 10.3** *Left* The squeezer optical bench, mounted on grouted tripod legs to obtain the needed beam height for squeezing injection into the interferometer. *Right* Small sample of the Squeezer Enclosure panelling.

**Fig. 10.4** Background displacement motion spectrum of the Squeezer table, as measured with the on-table accelerometers

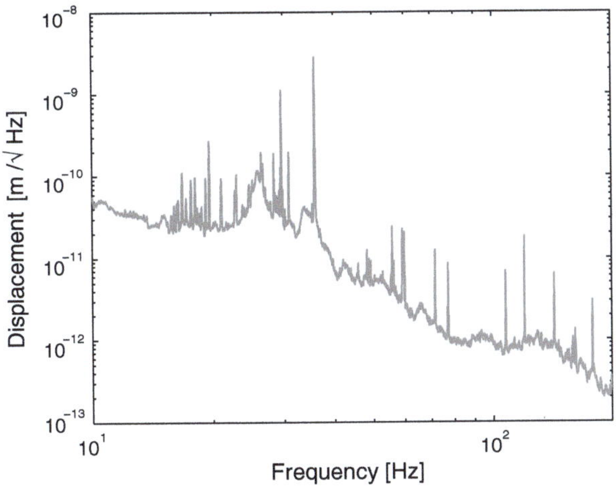

in thickness, were effective for their primary objectives of dust minimisation and stray-beam containment. However, it provided only limited air-current mitigation and minimal (at best) acoustic isolation.

## 10.4 Evidence of the Backscattered Light Mechanism

The backscattered light noise coupling mechanism has an identifiable signature in a readout spectrum. As backscattered light propagates to the readout photodetector, the light is imparted in phase with the displacement spectrum of objects in the beam path. Shown in Fig. 10.5, when applying a single frequency driven vibration, 'wing' structures were observed in the readout spectrum around that applied single frequency. The wing spectral shape can be compared with the motion spectrum of a test mass. There is similarity between the spectra, particularly the existence of sideband peaks at ~0.75 Hz, the pendulum frequency of the suspended test mass optics. This is evidence of the backscattered light noise coupling mechanism.
Further evidence of the backscattered light mechanism includes

i. when black glass beam dumps were placed into the squeezed light injection path with driven-vibration present, features in the interferometer readout spectrum were eliminated [10], and
ii. the disappearance of peaks in readout spectrum when the vibration unit was lifted slightly off the squeezer table while activated. This indicates the peaks were caused by the vibration motion, not other effects like stray magnetic fields of the active-vibration unit [10].

With evidence of the backscattered light mechanism present, the displacement measurements made to characterise the backscattered-light impact are now presented.

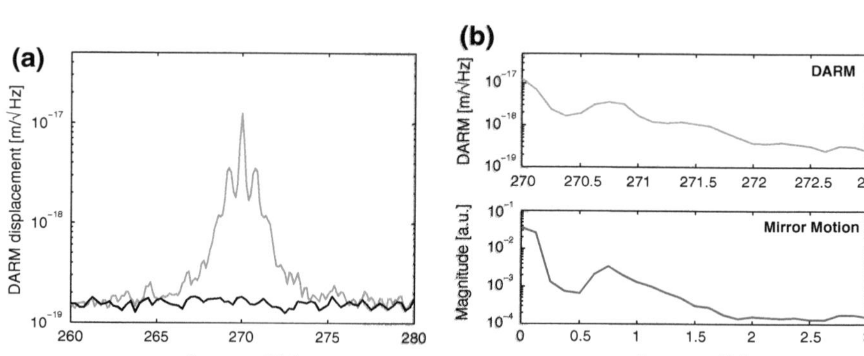


**Fig. 10.5** **a** Spectrum around a single frequency vibration injection at 270 Hz, showing 'wing' structures. **b** Comparison of wing structure spectrum (*top panel*) with a test mass position error-signal spectrum (*bottom panel*). The similarity between spectra indicates that the wing structure is a result of backscattered light sampling the motion of test masses before reaching the readout.

## 10.5 Large Displacement Measurement

The contribution of large displacements to backscatter noise arises from motion whose magnitude is comparable to the optical wavelength ($X \sim \lambda$), such as micro-seismic ground motion. Nonlinear noise upconversion [1] can result from motion of this scale, appearing as a "shelf" in the readout spectrum[2] [3]. This shelf-structure was not observed during the LIGO H1 Squeezed Light Injection Experiment when the interferometer and squeezer were operating under standard environmental conditions. However, we can induce a shelf-structure in the interferometer readout that will allow the backscattered light power level $P_s$ reaching the interferometer photodetector to be inferred. This technique was used previously to measure the backscatter influence of the interferometer Output Mode Cleaner cavities [11].

Using Eq. 10.4(i), the relative intensity noise spectrum for the backscatter signal $\mathrm{RIN}_s$ is given by

$$\mathrm{RIN}_s = 2\sqrt{\frac{P_s}{\bar{P}_c}} \cos\left(\frac{4\pi X_s}{\lambda}\right) \tag{10.7}$$

We intentionally modulate the backscatter optical path length with a known large-magnitude modulation. This is of the form $X = X_{dr} \sin(2\pi f_{dr} t)$, with large drive-magnitude $X_{dr}$ and drive frequency $f_{dr}$, we can induce a shelf-structure in the RIN spectrum. The RIN spectrum is now described as

$$\mathrm{RIN}_{dr} = 2\sqrt{\frac{P_s}{\bar{P}_c}} \cos\left(\mathrm{M} \sin 2\pi f_{dr} t\right) \tag{10.8}$$

where the modulation depth $M$ is given by

$$M = \frac{4\pi}{\lambda} X_{dr} \tag{10.9}$$

With known modulation depth $M$ and drive frequency $f_{dr}$, we can determine the backscatter-to-carrier power ratio ($P_s/\bar{P}_c$), and subsequently the backscatter power reaching the interferometer readout photodetector.

As shown in Fig. 10.1, a piezo-electric transducer (PZT) placed in the squeezing-injection path was used to modulate the optical path length. A sinusoidal modulation of known large modulation depth, $M = 173$ rad, and drive frequency, $f_{dr} = 1$ Hz, was applied. Shown in Fig. 10.6 is the induced shelf-structure in the measured interferometer RIN spectrum, and the modelled fit.

The data shown in Fig. 10.6 is corrected for the measured suppression of the interferometers control system loop [12]. A consistency check is the frequency of the shelf-structure cutoff, or *knee* frequency $f_{knee}$ [11]. This knee frequency should be equal to the drive frequency scaled by the modulation depth, or

[2] A derivation showing the shelf-structure characteristic can be found in Appendix C.

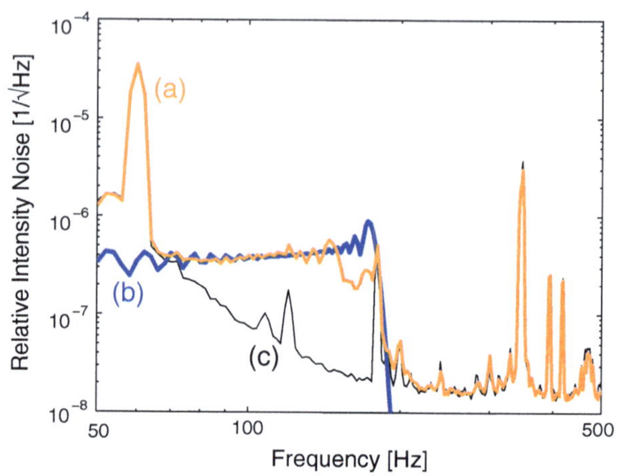


**Fig. 10.6** Large displacement measurement result **a** Relative-intensity noise of the interferometer readout, with induced shelf-structure from applied known backscatter pathlength modulation, and **b** modelled Eq. 10.8(i), using known parameters and fitting for $P_s/\bar{P}_c$. **c** Interferometer readout with no applied modulation. Reprinted with permission from [4], copyright (2013) by IOP Publishing

$$f_{\text{knee}} = M f_{\text{dr}} \tag{10.10}$$

The knee frequency for this measurement is $f_{\text{knee}} = 173$ Hz. This is consistent with the drive frequency scaled by the modulation depth.

From the modelled fit, the backscatter-to-carrier power ratio was found to be $P_s/\bar{P}_c = 1.7 \pm 0.2 \times 10^{-11}$. The backscatter power of the squeezer reaching the interferometer readout photodetector is therefore determined to be $P_s = 2.6 \pm 0.4 \times 10^{-13}$ W, calculated using the measured carrier power of 16.1 mW.

## 10.6 BSDF of the H1 Optical Parametric Oscillator

Using the above measurement of $P_s$, we make an estimate of the bi-directional scatter distribution function (BSDF) for the LIGO OPO. Similar to the process in Chap. 7, we again assume that the reflection occurs solely at the OPO nonlinear crystal.

The steps needed to estimate the BSDF are to first estimate of the power that is coupled into the reverse-propagating direction of the OPO, then estimate the forward-propagating scattered field just after the OPO. This provides us with the intrinsic isolation value $R_r$. Lastly correcting for the OPO cavity waist and OPO cavity finesse provides us with the BSDF estimate. Each step of the estimation is detailed next, with variables of interest presented in Table 10.1 and in Fig. 10.7.

- **Estimation of the power coupled into the reverse-propagating direction of the OPO**—The optical power incident on the squeezer table $P_{\text{s-inc}}$ was measured [13]. This is then corrected for the measured proportion of the carrier power in the AS port $\rho_{\text{C-AS}}$ [14] and the modematching of the field to the OPO, $\eta_{\text{sc-MM}}$. The modematching is $11 \pm 3\,\%$, estimated by an OPO path length scan measurement. Together, the power incident on the OPO that can be backscattered is

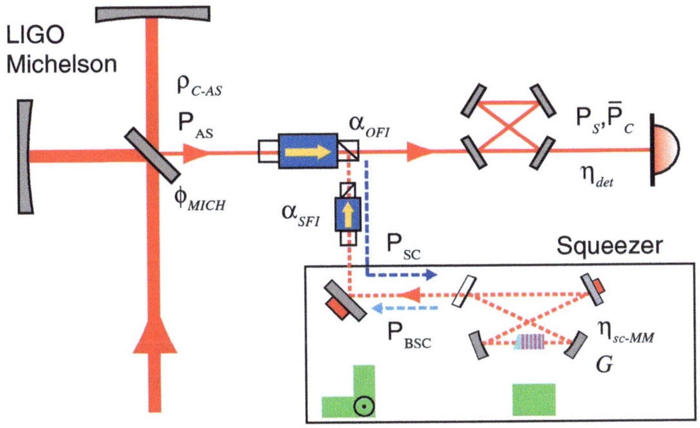


**Fig. 10.7** Simplified schematic showing variables of interest of the squeezed interferometer and backscattered-light path. The variables are defined in Table 10.1. This Figure is similar to Fig. 10.1

**Table 10.1** Variables for calculating the BSDF of the H1 Optical Parametric Oscillator.

| Variable | Symbol | Value |
|---|---|---|
| Squeezing detection efficiency | $\eta_{det}$ | 38 % |
| Classical OPO nonlinear gain | $G$ | 12 |
| Carrier/AS port power ratio | $\rho_{\text{C-AS}}$ | 1/5 |
| Total scatter power on OPO | $P_{\text{SC}}$ | 3 μW |
| Scatter modematching to OPO | $\eta_{\text{sc-MM}}$ | 11 % |

$$P_{\text{sc}} = P_{\text{s-inc}} \times \rho_{\text{C-AS}} \times \eta_{\text{sc-MM}} \tag{10.11}$$

- **Calculating the power leaving the OPO**—The measurement of backscatter power reaching the interferometer photodetector is firstly corrected for the squeezing detection efficiency $\eta_{det}$. This is then corrected for the known OPO parametric gain $G$, assuming the backscatter field is being deamplified. The forward reflecting field is thus

$$P_{\text{BSC}} = \frac{P_{\text{s}}}{\eta_{det}} \left( \frac{1+x}{1-x^2} \right)^{-2} \tag{10.12}$$

The intrinsic isolation is derived by dividing Eq. 10.12 by Eq. 10.11

$$R_{r-\text{H1}} = \frac{P_{\text{s}}}{P_{\text{SC}} \times \eta_{det}} \left( \frac{1+x}{1-x^2} \right)^{-2} \tag{10.13}$$

Calculation results in an estimated intrinsic isolation of $R_{r-\text{H1}} = 47 \pm 3\,\text{dB}$. Calculating the BSDF using the known OPO cavity finesse (Sect. 8.4) and beam waist size of 34 μm at the nonlinear crystal gives an estimated BSDF of $5 \pm 2 \times 10^{-4}$ str$^{-1}$.

The BSDF and intrinsic isolation values for the H1 OPO have uncertainty arising in particular from OPO mode matching and the OPO nonlinear gain. However, the BSDF value is better than the BSDF of the DB-OPO at ANU. This is credited to the use of superpolished and coated mirrors for the H1 OPO, reducing the surface scattering centres present.

## 10.7 Small Displacement Measurement

Small-displacement motions have displacement magnitudes much smaller than the optical wavelength ($\delta x_s \ll \lambda$). Within the gravitational-wave detection frequency band, backscattered light noise from small displacement motion contributions couple linearly to the readout noise spectrum [2, 3]. This linear coupling characteristic was verified in experiment,[3] and can also be used to infer the backscatter noise level below the sensitivity spectrum.

From Eq. 10.4(ii), the RIN spectral density of the background backscatter noise can be written as

$$\widetilde{\mathrm{RIN}}_s = 2\sqrt{\frac{P_s}{\bar{P}_c}}\frac{4\pi}{\lambda}\widetilde{\delta x}_{\mathrm{bg}} \tag{10.14}$$

where $\widetilde{\delta x}_{\mathrm{bg}}$ is the amplitude spectral density (ASD) of the background displacement motion. If we purposely drive the small-displacement motion $\widetilde{\delta x}_{\mathrm{dr}}$ of the squeezer table above the background motion ASD $\widetilde{\delta x}_{\mathrm{bg}}$, so that a noise signal $\mathrm{RIN}_{\text{s-in}}$ appears in the interferometer readout, we can then infer the background RIN of the backscatter $\mathrm{RIN}_{\text{s-bg}}$. Explicitly

$$\begin{aligned}\frac{\widetilde{\mathrm{RIN}}_{\text{s-bg}}}{\widetilde{\mathrm{RIN}}_{\text{s-dr}}} &= \frac{\widetilde{\delta x}_{\mathrm{bg}}}{\widetilde{\delta x}_{\mathrm{dr}}} \\ \rightarrow \widetilde{\mathrm{RIN}}_{\text{s-bg}} &= \widetilde{\mathrm{RIN}}_{\text{s-dr}} \times \frac{\widetilde{\delta x}_{\mathrm{bg}}}{\widetilde{\delta x}_{\mathrm{dr}}}\end{aligned} \tag{10.15}$$

To obtain a measure of the backscatter ASD from background table motion, the background ($\delta x_{\mathrm{bg}}$) and driven squeezer table motions ($\delta x_{\mathrm{in}}$) were measured by three orthogonally-mounted accelerometers (ACC—Wilcoxon Research, Model 731-207) on the squeezer optical table. In the driven measurement, small-magnitude squeezer table motion was induced by a piezo-shaker unit (SH—Piezo Systems Inc, Model IVBI-003-E). Figure 10.1 shows both equipment. Five single-frequency sinusoid shaker injections were made within the 50–300 Hz Fourier region. These injections were made at 75, 130, 155, 213, and 270 Hz.

[3]Shown in Appendix C.

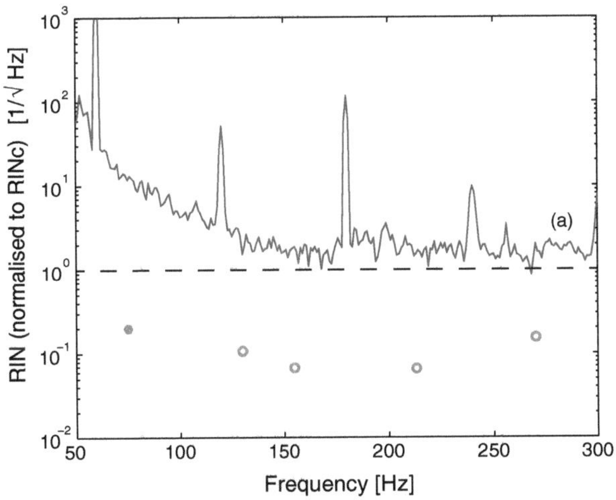


**Fig. 10.8** Small displacement measurement result—Inferred background backscatter RIN level for the *five* single-frequency small vibration injections. These injections were made within the 50–300 Hz detection region, chosen at 75, 130, 155, 213, and 270 Hz. The background backscatter RIN level is inferred to be greater than a factor of 10 below the interferometer readout RIN (trace **a**). This result shows that the backscatter impact of this installed squeezer does not detrimentally limit the interferometer sensitivity in the 50–300 Hz detection band

The resulting inferred noise level for injections is shown in Fig. 10.8. The inferred RIN level due to backscatter from small-displacement background motion is greater than a factor of 10 below the interferometer readout RIN. This result demonstrates that the backscatter impact of this installed squeezer does not detrimentally limit the interferometer sensitivity in the 50–300 Hz detection band. The injection at 75 Hz represents an upper limit to the backscatter noise, as no peak in the interferometer readout RIN was observed with the injection active.

## 10.8 Consistency Between Large Displacement and Small Displacement Measurements

To check the consistency between the large displacement and small displacement measurements, a backscatter RIN level can be calculated using Eq. 10.4(ii), $P_s$ obtained from the large displacement measurement and the ground motion spectrum shown in Fig. 10.4. This can then be compared to the inferred points from the small displacement measurement. The result of this consistency calculation is shown in Fig. 10.9.

The calculated backscatter RIN level agrees with the inferred points from the small displacement measurement. This indicates that the large displacement and small displacement measurement results are consistent.

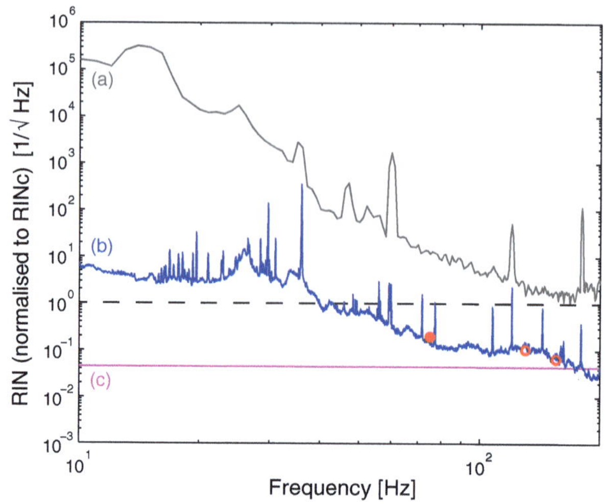


**Fig. 10.9** Comparison between backscatter RIN results for the two displacement measurements. *Circles* Inferred backscatter RIN level from the small displacement measurement. Trace **a** is the interferometer readout RIN. Trace **b** is the result of Eq. 10.4(ii) using $P_s$ from the large displacement measurement, the ground motion shown in Fig. 10.4 and carrier power $\bar{P}_c$. Trace **c** is the clearance level for Advanced LIGO readout, discussed in Sect. 10.9

## 10.9 Backscatter Improvements for Future Gravitational-Wave Detectors with Squeezing

The measurements reported have confirmed that backscattered light impacts represents a technical challenge, not a fundamental detriment to interferometer readout. However, future gravitational-wave detectors such as Advanced LIGO will place stricter tolerances to backscatter, due to increased gravitational-wave sensitivity, increased carrier power in the AS port, as well as higher detection efficiencies for applied squeezing [15]. We would like a clearance level of 10 below the RIN of Advanced LIGO readout. This clearance level is shown in Fig. 10.9. The backscatter RIN for the LIGO H1 squeezer is ~200 times above the Advanced LIGO clearance level below 50 Hz, and ~7 times above the clearance level from 50 Hz.

Further mitigation of the impact of backscatter is needed in future gravitational-wave detectors. The backscatter RIN equation (Eq. 10.4) highlights the mitigation options:

- **Reducing the beam path displacement motion** $X$—Technology options for reducing the beam path displacement motion are currently available. For second generation detectors, environmental isolation with acoustic enclosures and specialised optical table tripods can be deployed to gain the required reduction factor of 200. For third generation detectors and beyond, active seismic-isolation platforms such as those being installed for the Advanced LIGO Output Mode Cleaners [16], and in-vacuum operation are possible isolation options.

- **Reducing the backscatter power** $P_S$—Reducing the backscatter power will involve improving the optics in the injection and AS port beam paths. As an example, increasing the isolation of both the Squeezed Faraday isolator and Output Faraday isolator to a more nominal 35 dB attenuation will provide a factor of $\sim$20 improvement alone, relaxing the motion-isolation required. Development work into improved optics and optical coatings will be beneficial here.

Reductions in the beam path displacement motion and the backscatter power reaching the interferometer readout photodetector are obtainable, to reduce the impact of backscattered light in Advanced LIGO and beyond.

## 10.10 Summary

It is shown that from a backscattered-light perspective, the presence of a squeezer represents a technical challenge, not a fundamental detriment, to the sensitivity of a gravitational-wave detector. The large displacement measurement inferred the backscatter light power incident on the readout photodetector to be $P_S = 2.6 \pm 0.4 \times 10^{-13}$ W. The bi-directional scatter distribution function of the LIGO H1 Optical Parametric Oscillator was estimated to be $5 \pm 2 \times 10^{-4}$ str$^{-1}$, a value better than the BSDF of the DB-OPO at ANU, credited to the use of the superpolished cavity mirrors. The small displacement measurement inferred the backscatter RIN level to be at least a factor of 10 below the interferometer readout curve, demonstrating the backscatter noise from the installed squeezer did not detrimentally limit the LIGO interferometer sensitivity. Suggested improvements to help mitigate the backscattered light impact of an installed squeezer in future gravitational-wave detectors include technology to reduce squeezer table motion and improved optical components for isolation.

## References


1. J.-Y. Vinet, V. Brisson, S. Braccini, I. Ferrante, L. Pinard, F. Bondu, E. Tournié, Scattered light noise in gravitational wave interferometric detectors: a statistical approach. Phys. Rev. D **56**, 6085–6095 (1997)
2. J.-Y. Vinet, V. Brisson, S. Braccini, Scattered light noise in gravitational wave interferometric detectors: coherent effects. Phys. Rev. D **54**, 1276–1286 (1996)
3. D.J. Ottaway, P. Fritschel, S.J. Waldman, Impact of upconverted scattered light on advanced interferometric gravitational wave detectors. Opt. Express **20**, 8329 (2012)
4. S.S.Y. Chua, S. Dwyer, L. Barsotti, D. Sigg, R.M.S. Schofield, V.V. Frolov, K. Kawabe,M. Evans, G.D. Meadors, M. Factourovich, R. Gustafson, N. Smith-Lefebvre, C. Vorvick,M. Landry, A. Khalaidovski, M.S. Stefszky, C.M. Mow-Lowry, B.C. Buchler, D.A. Shaddock,P.K. Lam, R. Schnabel, N. Mavalvala, D.E. McClelland, Class. Quantum Grav. **31**,035017 (2014)

5. E. Flanagan, K.S. Thorne, Noise due to backscatter off baffles, the nearby wall, and objects at the far end of the beam tube; and recommended actions (1994). LIGO-T940063-00-R
6. E. Flanagan, K.S. Thorne, Scattered-light noise for ligo (1995). LIGO-T950132-00-R
7. P. Fritschel, M.E. Zucker, Wide-angle scatter from LIGO arm cavities (2007). LIGO-T070089-05-D
8. L. Barsotti, Power from IFO to SQZ table. Hanford iLog, 19 Oct 2011
9. Lisa Barsotti, aLIGO Faraday: T = 0.96, Isolation 40 dB, but only 23 dB backscatter isolation. Hanford iLog, 9 June 2011
10. R. Schofield, The impact of squeezing on vibration coupling to H1. Hanford iLog, 23 Oct 2011
11. T. Fricke, Homodyne detection for interferometric gravitational-wave detectors. Ph.D. thesis, Graduate Faculty, Louisiana State University, 2011
12. S. Chua, H1OLG. Hanford iLog, 28 Nov 2011
13. S. Chua, Power returning back from HAM4 + Polarization. Hanford iLog, 3 Dec 2011
14. L. Barsotti, Sideband/carrier power @ 8W. Hanford iLog, 29 Nov 2011
15. R. Abbott, R. Adhikari, S. Ballmer, L. Barsotti, M. Evans, P. Fritschel, V. Frolov, G. Mueller, B. Slagmolen, S. Waldman, Advanced LIGO length sensing and control final design (2010). LIGO-T1000298-T
16. N.A. Robertson, M. Barton, D. Bridges, J. Heefner, M. Meyer, J. Romie, C. Torrie. OMC suspension final design document (2009). LIGO-T0900060-v1

# Chapter 11
# Results Summary, Recommendations and Future Work

**Abstract** The technology to generate and implement quantum squeezed states to enhance gravitational-wave detector sensitivity has been the focus of this thesis. This chapter presents a brief summary of the findings from the work presented in this thesis. From these results, a number of future investigations are proposed.

## 11.1 Summary of Squeezed-Light Source Apparatus Development Results

- The ANU-developed Optical Parametric Oscillator (OPO) and its design reasoning was presented. Measurements of squeezing demonstrated the first observation of greater than 10 dB squeezing across the audio gravitational-wave detection band ($11.6 \pm 0.4$ dB above 200 Hz), and 5900 s of continuously controlled squeezing at $7.5 \pm 0.5$ dB. This result was characterised in terms of measurement efficiencies and squeezing ellipse phase noise, resulting in an inferred squeezing magnitude level exiting the OPO of $17.0 \pm 0.3$ dB.
- A modified coherent sideband locking scheme used for squeezing ellipse phase control was introduced. This modified technique led to an increase of unity-gain bandwidths of control loops, from 10 s/100 s of Hz to 10 s of kHz.
- The tolerance of a bow-tie OPO to backscattered light was modelled, and experimentally verified using the homodyne detector. An intrinsic isolation of $41 \pm 2$ dB was measured, at least comparable to off-the-shelf Faraday isolators, and an additional suppression of approximately 40 dB over an equivalent standing-wave cavity. The bi-directional scatter distribution function (BSDF) of the OPO was determined to be $1.7 \pm 0.3 \times 10^{-3}$.


S.S.Y. Chua, *Quantum Enhancement of a 4 km Laser Interferometer Gravitational-Wave Detector*, Springer Theses,
DOI 10.1007/978-3-319-17686-4_11

## 11.2 Summary of LIGO Squeezed Light Injection Experiment Results

- The LIGO Squeezed light Injection Experiment was introduced. This included the project background, the Enhanced LIGO interferometer with the hardware modifications made for this Experiment, and the squeezed light injection optical path. The squeezed light source and the ANU-contributed OPO were presented.
- The first observation of squeezing-enhancement of a 4 km gravitational-wave detector was presented. The squeezing surpassing the quantum shot noise by up to $2.15 \pm 0.05$ dB down to 150 Hz detection frequency. Above 250 Hz, this result represents the best sensitivity in a gravitational-wave detector yet achieved. The presence of a squeezed-light source added no noise to low frequency sensitivity. The capability of squeezing to recover sensitivity using lower interferometer input power was also demonstrated.
- The parameters of the LIGO Squeezed Light Injection Experiment that impact the observed squeezing-enhancement were identified and quantified. The results of squeezing ellipse phase noise, originating from contributions of the squeezer apparatus, interferometer control sidebands, and inferred beam pointing phase noise were reported. The squeezing detection efficiency was measured, then verified using two characterisation methods. The expected squeezing enhancement was inferred from the squeezing level generated by the OPO, correcting for the squeezing detection efficiency and squeezing ellipse phase noise. The measured squeezing enhancement and the expected squeezing enhancement were found to be in good agreement.
- From a backscattered-light perspective, the presence of a squeezer represents a technical challenge, not a fundamental detriment, to the sensitivity of a gravitational-wave detector. The results of two different experiments were presented, resulting in the inferring of the backscatter RIN level to be a factor of 10 below the interferometer readout RIN, and the backscatter light power incident on the readout photodetector to be $P_s = 2.6 \pm 0.4 \times 10^{-13}$ W. The BSDF of the LIGO Optical Parametric Oscillator was then estimated to be $5 \pm 2 \times 10^{-4}$ str$^{-1}$, a value better than the BSDF of the DB-OPO at ANU. This is credited to the use of the superpolished cavity mirrors in the H1-OPO.

## 11.3 Summary of Recommended Development Work Towards Squeezed-Future Gravitational-Wave Detectors

### *11.3.1 Improving Squeezing Detection Efficiency*

- **Component optical loss**—Improving optical loss in components of the squeezing injection path will be a considerable research/ development undertaking.

Increasing the transmission efficiency of critical components such as Faraday isolators, and increasing the Output Mode Cleaner (OMC) throughput will be needed in order to access higher levels of applied squeezing.

- **Spatial Modematching to the OMC**—The first improvement is to increase the OMC optical path length scan rate. This then allows many more iterations to assess the modematching of the squeezed beam to the OMC cavity mode. The improvement is part of Advanced LIGO OMC design specifications [1].

  The second improvement is to replace the fixed modematching lens system with a tunable modematching system as part of the squeezing injection path. A tunable modematching system potentially offers the ability to precisely modematch the squeezed beam to the OMC, and the ability to in-situ alter the modematching to compensate for changes from thermal heating of the full interferometer.

### *11.3.2 Reducing Sources of Squeezing Ellipse Phase Noise*

- **From the squeezer apparatus**—Phase noise on the squeezer table is dominated by path length fluctuations of optical beams. Mechanisms for these fluctuations include acoustic noise and air currents. Possible methods to improve the phase noise include enclosing critical beam paths, covering the squeezer apparatus with airtight enclosures, and moving the squeezer OPO into the interferometer vacuum envelope.
- **From the interferometer control sidebands**—Minimising the proportion of contrast defect field power in the interferometer readout will minimise the interferometer sidebands contribution to phase noise.
- **From beam pointing fluctuations**—An active alignment scheme should be installed to minimise beam pointing fluctuations. Readout of the beam alignment can be achieved using quadrant photodiodes measuring the relative beam pointing between the interferometer carrier field and squeezing coherent control sidebands. Actuation on the squeezed beam can be fed back to the two PZT steering mirrors on the squeezer table (Sect. 8.3).

### *11.3.3 Reducing Backscattered Light*

There are several avenues to pursue in order to reduce the backscattered light impact.

- **Reducing the beam path displacement motion** $X$—Technology options for reducing the beam path displacement motion are currently available. For second generation detectors, environmental isolation with acoustic enclosures and specialised optical table tripods can be deployed to gain the required reduction factor of 200. For third generation detectors and beyond, active seismic-isolation

platforms such as those being installed for the Advanced LIGO Output Mode Cleaners [2], and in-vacuum operation are possible isolation options.
- **Reducing the backscatter power** $P_s$—Reducing the backscatter power will involve improving the optics in the injection and AS port beam paths. As an example, increasing the isolation of both the Squeezed Faraday isolator and Output Faraday isolator to a more nominal 35 dB attenuation will provide a factor of $\sim$20 improvement alone, relaxing the motion-isolation required. Development work into improved optics and optical coatings will be beneficial here.

## 11.4 Future Work

In addition to the development recommendations in the previous section, these areas of future research are proposed:

- In-vacuum squeezed light sources—Installing, characterising and operating a squeezed light source in-vacuum raises new challenges. These technical challenges include the need for new procedures to optimise nonlinear crystal position and phase matching temperature, as well as work in specifying experiment parameter tolerances.
- All-glass Optical Parametric Oscillator—An all-glass Optical Parametric Oscillator is the next engineering development to the OPO design. An optical cavity with glass-bonded mirrors will provide stability and robustness for inclusion into future gravitational-wave detectors, particularly when squeezed light sources are housed inside vacuum chambers.
- Phase matching temperature readout—Current readout of the temperature for phase matching rely on temperature sensors away from the optical nonlinear interaction zone within the crystal. This can lead to drifts in temperature actuation away from optimum phase matching. Developing a method to measure the phase matching temperature at the nonlinear interaction region will make long-term squeezer operation more stable.
- Integration with filter cavities—In order to obtain frequency-dependent squeezed states, current knowledge advises that the squeezed state needs to be integrated with optical filter cavities (Chap. 4, and [3]). The demonstration of optical filter cavities providing squeezing ellipse rotation in the audio detection band is yet to be achieved. Research is on going into the design and construction of suitable filter cavities for frequency dependent squeezing (for example [4]).

## References

1. K. Arai, S. Barnum, P. Fritschel, J. Lewis, S. Waldman, Output mode cleaner design (2013). LIGO-T1000276-v5
2. N.A. Robertson, M. Barton, D. Bridges, J. Heefner, M. Meyer, J. Romie, C. Torrie, OMC suspension final design document (2009). LIGO-T0900060-v1

3. H.J. Kimble, Y. Levin, A.B. Matsko, K.S. Thorne, S.P. Vyatchanin, Conversion of conventional gravitational-wave interferometers into quantum nondemolition interferometers by modifying their input and/or output optics. Phys. Rev. D **65**, 022002 (2002)
4. P. Kwee, L. Barsotti, M. Evans, M. Zucker, P. Fritschel, N. Mavalvala, Proposal for a filter cavity experiment at last (2012). LIGO-T1200024-v3

# Appendix A
# Quantum Noise and Squeezing in Dual-Recycled Michelson Interferometers

**Abstract** This appendix adds to the theory covered in Chap. 4—*Quantum noise in Gravitational-wave Detectors and Applied Squeezed States*, reviewing the quantum noise equations for Dual-recycled Michelson Interferometers, and the improvements from injected squeezed states. These equations are from the works of Buonanno and Chen [1, 2].

## A.1 Quantum Noise in a Lossless Dual-Recycled Michelson Interferometer

The power spectral density of a lossless Dual-recycled Michelson with arm cavities is given by [1]

$$S^{\zeta}_{h-DR} = \mathscr{Q} \times [(C_{11}\sin\zeta + C_{21}\cos\zeta)^2 + (C_{21}\sin\zeta + C_{22}\cos\zeta)^2] \tag{A.1}$$

with

$$\mathscr{Q} = \frac{(h^{\mathrm{I}}_{SQL})^2}{2\mathscr{K}^{\mathrm{I}}\tau^2|D_1\sin\zeta + D_2\cos\zeta|^2} \tag{A.2}$$

and

$$D_1 = -(1+\rho e^{2i(\beta+\Theta)})\sin\phi \tag{A.3}$$
$$D_2 = -(-1+\rho e^{2i(\beta+\Theta)})\cos\phi \tag{A.4}$$
$$C_{11} = (1+\rho^2)\left(\cos 2\phi + \frac{\mathscr{K}^{\mathrm{I}}}{2}\sin 2\phi\right) - 2\rho\cos(2\beta+2\Theta) \tag{A.5}$$
$$C_{12} = -\tau^2(\sin 2\phi + \mathscr{K}^{\mathrm{I}}\sin^2\phi) \tag{A.6}$$
$$C_{21} = \tau^2(\sin 2\phi - \mathscr{K}^{\mathrm{I}}\cos^2\phi) \tag{A.7}$$
$$C_{22} = C_{11} \tag{A.8}$$

Recall $\zeta$ being the detection angle, and $\mathscr{K}^{\mathrm{I}}$ and $h^{\mathrm{I}}_{SQL}$ being the modified parameters from the presence of arm cavities. The other variables are the amplitude reflectivity



S.S.Y. Chua, *Quantum Enhancement of a 4 km Laser Interferometer Gravitational-Wave Detector*, Springer Theses,
DOI 10.1007/978-3-319-17686-4

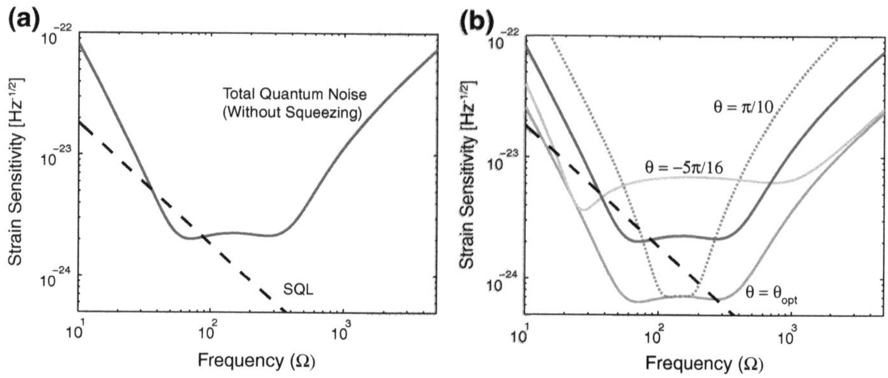


**Fig. A.1** **a** Strain sensitivity of a lossless Dual-recycled Michelson and the (free-mass) SQL. The trace shown is with a non-zero SR cavity detuning offset of $\delta\phi = 0.2$ rad. The quantum noise beats the SQL within a narrow frequency band, a feature gained due to signal recycling. **b** The enhanced strain sensitivity with squeezing at various squeezing angles, superimposed with the non-squeezed strain trace

$\rho$ and transmittivity $\tau$ of the signal-recycling mirror (SRM), the SR cavity detuning $\phi$ and the SR cavity single-pass phase shift $\Theta$. Figure A.1 shows the quantum noise limited strain sensitivity $\sqrt{S_{h-DR}}$, utilising the values in Table A.1, as well as the SQL $h^{\mathrm{I}}_{SQL}$. By choosing an appropriate SR cavity detuning, the quantum noise can be lower than the SQL at certain frequencies [1].

### *A.1.1 Lossless Dual-Recycled Michelson with Squeezing*

The injection of squeezing into lossless Dual-recycled Michelson gravitational-wave detectors have been analysed in [2–4]. The power spectral density of the quantum noise with squeezing is given by [2]

$$S^{\zeta}_{\mathrm{h-DRSQZ}} = \mathscr{Q} \times [e^{-2R}(C^{\theta}_{11}\sin\zeta + C^{\theta}_{21}\cos\zeta)^2 + e^{2R}(C^{\theta}_{21}\sin\zeta + C^{\theta}_{22}\cos\zeta)^2] \tag{A.9}$$

The C-coefficients are now a combination of the original C-coefficients, weighted by the projection with the squeezing angle $\theta$, namely

$$C^{\theta}_{11} = C_{11}\cos\theta + C_{12}\sin\theta \tag{A.10}$$
$$C^{\theta}_{12} = C_{12}\cos\theta - C_{11}\sin\theta \tag{A.11}$$
$$C^{\theta}_{21} = C_{21}\cos\theta + C_{22}\sin\theta \tag{A.12}$$
$$C^{\theta}_{22} = C_{21}\cos\theta - C_{22}\sin\theta \tag{A.13}$$

**Table A.1** Table of values for quantum noise calculations of the Dual-recycled Michelson

| Parameter | Symbol | Value | Units |
|---|---|---|---|
| Laser frequency | $\omega_0$ | $1.77 \times 10^{15}$ | rad/s |
| Detection Fourier frequency | $\Omega$ | 10–10,000 | Hz |
| Detection phase | $\zeta$ | $\pi/2$ | rad |
| Photodetection loss | $\lambda_{PD}$ | 0.1 | – |
| Laser power at beamsplitter | $I_0$ | 5067 | W |
| Mirror mass | $m$ | 40 | kg |
| ITM transmission | $T$ | 0.014 | – |
| Arm length | $L$ | 3995 | m |
| Arm cavity half-bandwidth | $\gamma$ | $Tc/(4L)$ $\approx 2\pi \times 40$ Hz | rad |
| Arm cavity round trip loss (per cavity) | $\lambda_{AC}$ | $38 \times 10^{-6}$ | – |
| Arm cavity loss coefficient | $\varepsilon$ | $2\lambda_{AC}/T$ | – |
| SRM transmission amplitude | $\tau$ | $\sqrt{0.2}$ | – |
| SRM reflection amplitude | $\rho$ | $\sqrt{1-\tau^2-\lambda_{SR}}$ | – |
| SR cavity loss | $\lambda_{SR}$ | $2 \times 10^{-3}$ | – |
| SR cavity length | $l$ | 55 | m |
| SR cavity detuning offset | $\delta\phi$ | 0.2[a], 0 | rad |
| SR cavity detuning | $\phi$ | $(\pi - \delta\phi)/2$ | rad |
| SR cavity single-pass phase shift | $\Theta$ | $(l\Omega/c)_{\text{mod}2\pi}$ | rad |

[a]Only for the lossless Dual-recycled Michelson to illustrate sub-SQL performance

Figure A.1b shows the enhanced strain sensitivity with squeezing at various squeezing angles. With optimal frequency-dependent squeezing, the resulting strain sensitivity is the non-squeezed strain sensitivity scaled by a factor of $\sqrt{e^{-2R}} = e^{-R}$ [4].

### *A.1.2 Dual-Recycled Michelson with Optical Losses*

Figure A.2 shows the loss (and hence vacuum-entrance) points for the generalised Dual-recycled Michelson with optical loss. Combined with the recycling of the output noise (along with the signal), this greatly complicates the calculation of quantum noise at the interferometer output, with more coefficients (indicated with superscript $L$) to be accounted for.

From Eq. 5.13 of [1], the quantum-noise-limited strain sensitivity is given by

$$h_{DR-L} = \sqrt{S^{\zeta}_{h-DRL}} \tag{A.14}$$

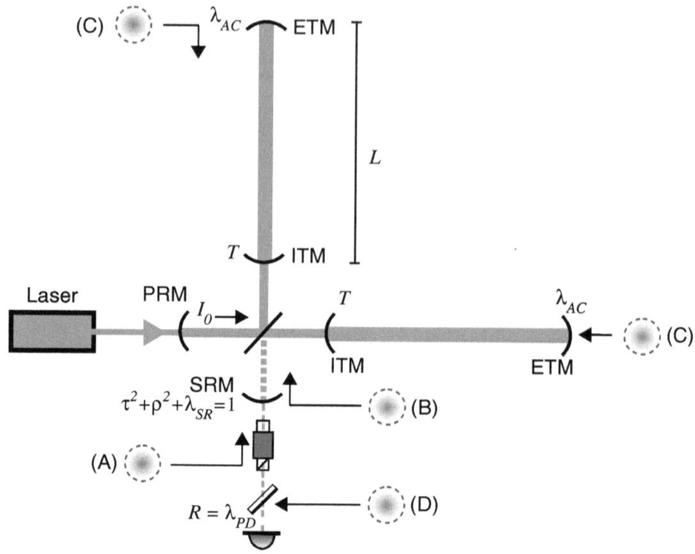


**Fig. A.2** Dual-recycled Michelson with Fabry-Perot arm cavities showing the entrance of vacuum fluctuations. *PRM* Power-recycling mirror; *ITM* Input test mass; *ETM* End test mass; *SRM* Signal-recycling mirror; λ Optical loss

with

$$
\begin{aligned}
S^{\zeta}_{h-DRL} = \mathscr{Q}^L \times [\, & \underbrace{|C^L_{11}\sin\zeta + C^L_{21}\cos\zeta|^2 + |C^L_{12}\sin\zeta + C^L_{22}\cos\zeta|^2}_{\text{A}} \\
& + \underbrace{|P^L_{11}\sin\zeta + P^L_{21}\cos\zeta|^2 + |P^L_{12}\sin\zeta + P^L_{22}\cos\zeta|^2}_{\text{B}} \\
& + \underbrace{|N^L_{11}\sin\zeta + N^L_{21}\cos\zeta|^2 + |N^L_{12}\sin\zeta + N^L_{22}\cos\zeta|^2}_{\text{C}} \\
& + \underbrace{|Q^L_{11}\sin\zeta + Q^L_{21}\cos\zeta|^2 + |Q^L_{12}\sin\zeta + Q^L_{22}\cos\zeta|^2]}_{\text{D}}
\end{aligned}
\tag{A.15}
$$

The coefficients in the above equation correspond to Eqs. 5.8–5.12 in [1], reproduced below for completeness. The quotient term is

$$
\mathscr{Q}^L = \frac{(h^{\mathrm{I}}_{SQL})^2}{2\mathscr{K}^{\mathrm{I}}\tau^2|D^L_1\sin\zeta + D^L_2\cos\zeta|^2}
\tag{A.16}
$$

with denominator D-terms

$$
\begin{aligned}
D^L_1 = \sqrt{1-\lambda_{PD}}\Big\{ & -(1+\rho e^{2i\beta^{\mathrm{I}}})\sin\phi \\
& + \frac{\varepsilon}{4}[3+\rho+2\rho e^{4i\beta^{\mathrm{I}}} + e^{2i\beta^{\mathrm{I}}}(1+5\rho)]\sin\phi + \frac{\lambda_{SR}}{2}e^{2i\beta^{\mathrm{I}}}\rho\sin\phi\Big\}
\end{aligned}
\tag{A.17}
$$

$$D_2^L = \sqrt{1-\lambda_{PD}}\Big\{(1-\rho e^{2i\beta^{\mathrm{I}}})\cos\phi$$
$$+\ \frac{\varepsilon}{4}[-3+\rho+2\rho e^{4i\beta^{\mathrm{I}}} - e^{2i\beta^{\mathrm{I}}}(1-5\rho)]\cos\phi + \frac{\lambda_{SR}}{2}e^{2i\beta^{\mathrm{I}}}\rho\cos\phi\Big\} \tag{A.18}$$

The C-coefficients for the vacuum noise entering the AS port (Noise A of Fig. A.2) are

$$C_{11}^L = \sqrt{1-\lambda_{PD}}\Bigg\{(1+\rho^2)\left(\cos 2\phi + \frac{\mathscr{K}^{\mathrm{I}}}{2}\sin 2\phi\right) - 2\rho\cos 2\beta^{\mathrm{I}}$$
$$-\ \frac{\varepsilon}{4}\Big[-2(1+e^{2i\beta^{\mathrm{I}}})^2\rho + 4(1+\rho^2)\cos^2(\beta^{\mathrm{I}})\cos 2\phi$$
$$+(3+e^{2i\beta^{\mathrm{I}}})\mathscr{K}^{\mathrm{I}}(1+\rho^2)\sin 2\phi\Big]$$
$$+\ \lambda_{SR}\left[e^{2i\beta^{\mathrm{I}}}\rho - \frac{1}{2}(1+\rho^2)\left(\cos 2\phi + \frac{\mathscr{K}^{\mathrm{I}}}{2}\sin 2\phi\right)\right]\Bigg\} \tag{A.19}$$

$$C_{12}^L = \sqrt{1-\lambda_{PD}}\tau^2\Big(-(\sin 2\phi + \mathscr{K}^{\mathrm{I}}\sin^2\phi) + \frac{\varepsilon}{2}\sin\phi[(3+e^{2i\beta^{\mathrm{I}}})\mathscr{K}^{\mathrm{I}}\sin\phi + 4\cos^2(\beta^{\mathrm{I}})\cos\phi]$$
$$+\ \frac{\lambda_{SR}}{2}(\sin 2\phi + \mathscr{K}^{\mathrm{I}}\sin^2\phi)\Big) \tag{A.20}$$

$$C_{21}^L = \sqrt{1-\lambda_{PD}}\tau^2\Big((\sin 2\phi - \mathscr{K}^{\mathrm{I}}\cos^2\phi)$$
$$+\ \frac{\varepsilon}{2}\cos\phi[(3+e^{2i\beta^{\mathrm{I}}})\mathscr{K}^{\mathrm{I}}\cos\phi - 4\cos^2(\beta^{\mathrm{I}})\sin\phi] + \frac{\lambda_{SR}}{2}(-\sin 2\phi + \mathscr{K}^{\mathrm{I}}\cos^2\phi)\Big) \tag{A.21}$$

$$C_{22}^L = C_{11}^L \tag{A.22}$$

The P-coefficients for the vacuum noise entering due to loss in the SR cavity (Noise B) are

$$P_{11}^L = \frac{1}{2}\sqrt{1-\lambda_{PD}}\sqrt{\lambda_{SR}}\tau(-2\rho e^{2i\beta^{\mathrm{I}}} + 2\cos 2\phi + \mathscr{K}^{\mathrm{I}}\sin 2\phi) \tag{A.23}$$
$$P_{12}^L = -\sqrt{1-\lambda_{PD}}\sqrt{\lambda_{SR}}\tau\sin\phi(2\cos\phi + \mathscr{K}^{\mathrm{I}}\sin\phi) \tag{A.24}$$
$$P_{21}^L = \sqrt{1-\lambda_{PD}}\sqrt{\lambda_{SR}}\tau\cos\phi(2\sin\phi - \mathscr{K}^{\mathrm{I}}\cos\phi) \tag{A.25}$$
$$P_{22}^L = P_{11}^L \tag{A.26}$$

The N-coefficients for the vacuum noise entering due to loss in the arm cavities (Noise C) are

$$N_{11}^L = \sqrt{1-\lambda_{PD}}\sqrt{\frac{\varepsilon}{2}}\tau\Big[\mathscr{K}^{\mathrm{I}}(1+\rho e^{2i\beta^{\mathrm{I}}})\sin\phi$$
$$+\ 2\cos\beta^{\mathrm{I}}(e^{-i\beta^{\mathrm{I}}}\cos\phi - \rho e^{i\beta^{\mathrm{I}}}(\cos\phi + \mathscr{K}^{\mathrm{I}}\sin\phi))\Big] \tag{A.27}$$

$$N_{12}^L = -\sqrt{1-\lambda_{PD}}\sqrt{2\varepsilon}\tau(e^{-i\beta^{\mathrm{I}}} + \rho e^{i\beta^{\mathrm{I}}})\cos\beta^{\mathrm{I}}\sin\phi \tag{A.28}$$

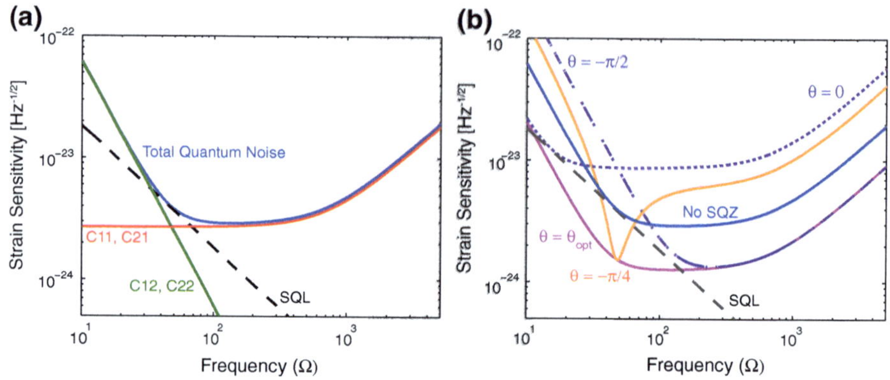


**Fig. A.3** **a** Strain sensitivity of a Dual-recycled Michelson with optical losses and the contribution of the dominant C-coefficient terms. These plots use the values in Table A.1 at the end of this Appendix, with zero SR cavity detuning offset $\delta\phi = 0$. The total quantum noise is the form plotted for the quantum noise contribution in GwinC v3 [5] (see Fig. 4.6 for comparison). **b** Strain sensitivity with injected squeezing at various squeezing angles

$$N^L_{21} = \sqrt{1-\lambda_{PD}}\sqrt{\frac{\varepsilon}{2}}\tau[-\mathscr{K}^{\mathrm{I}}(1+\rho)\cos\phi + 2\cos^2\beta^{\mathrm{I}}(e^{-i\beta^{\mathrm{I}}} + \rho e^{i\beta^{\mathrm{I}}})\sin\phi] \tag{A.29}$$

$$N^L_{22} = -\sqrt{1-\lambda_{PD}}\sqrt{2\varepsilon}\tau(-e^{-i\beta^{\mathrm{I}}} + \rho e^{i\beta^{\mathrm{I}}})\cos\beta^{\mathrm{I}}\cos\phi \tag{A.30}$$

Lastly, the Q-coefficients for the vacuum noise entering due to imperfect photodetection (Noise D) are

$$\begin{aligned} Q^L_{11} = Q^L_{22} = \sqrt{\lambda_{PD}}\Big\{& e^{-2i\beta^{\mathrm{I}}} + \rho^2 e^{2i\beta^{\mathrm{I}}} - \rho(2\cos 2\phi + \mathscr{K}^{\mathrm{I}}\sin 2\phi) + \frac{\varepsilon}{2}\rho\Big[e^{-2i\beta^{\mathrm{I}}}\cos 2\phi \\ & + e^{2i\beta^{\mathrm{I}}}(-2\rho - 2\rho\cos 2\beta^{\mathrm{I}} + \cos 2\phi + \mathscr{K}^{\mathrm{I}}\sin 2\phi) + 2\cos 2\phi + 3\mathscr{K}^{\mathrm{I}}\sin 2\phi\Big] \\ & - \frac{\lambda_{SR}}{2}\rho(2\rho e^{2i\beta^{\mathrm{I}}} - 2\cos 2\phi - \mathscr{K}^{\mathrm{I}}\sin 2\phi)\Big\} \end{aligned} \tag{A.31}$$

$$Q^L_{12} = Q^L_{21} = 0 \tag{A.32}$$

These coefficients are simplified by the sideband-phase gain in the SR cavity being set to zero ($\Phi = 0$). Figure A.3 shows the strain sensitivity of a Dual-recycled Michelson with optical losses, calculated using the values in Table A.1 at the end of this Appendix, with zero SR cavity detuning offset $\delta\phi = 0$. Also plotted are traces of the contribution of the AS port vacuum fluctuations (C-coefficient terms) that dominate the total quantum noise. This is the form plotted for the quantum noise contribution in GWINC v3 [5] (see Fig. 4.6 for comparison) (Fig. A.4).

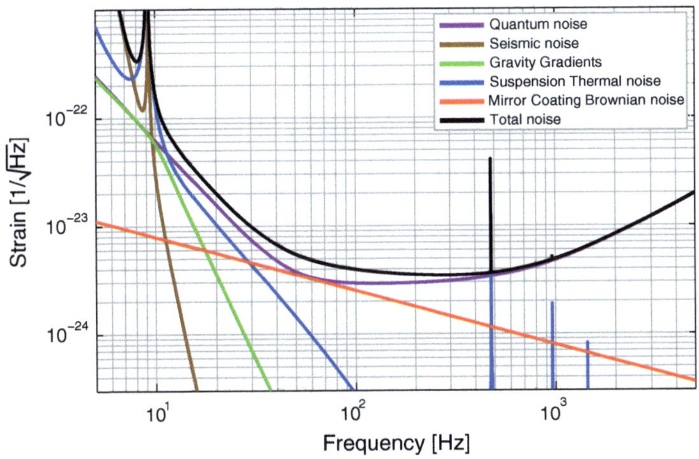


**Fig. A.4** Figure 4.6 repeated—the predicted noise budget of Advanced LIGO, showing the contributions of the various limiting noise sources. Plot made using GwinC v3 [5]

## A.2 Dual-Recycled Michelson with Losses and Squeezing

In a similar way to the derivation in the lossless case, the power spectral density of the dual-recycled Michelson with losses and injected squeezing can be derived, given by [2]

$$
\begin{aligned}
S^{\zeta}_{h-DRLSQZ} = \mathscr{Q}^L \times \Big[ & e^{-2R} |C^{L\theta}_{11} \sin\zeta + C^{L\theta}_{21} \cos\zeta|^2 + e^{2R} |C^{L\theta}_{21} \sin\zeta + C^{L\theta}_{22} \cos\zeta|^2 \\
& + |P^L_{11} \sin\zeta + P^L_{21} \cos\zeta|^2 + |P^L_{21} \sin\zeta + P^L_{22} \cos\zeta|^2 \\
& + |N^L_{11} \sin\zeta + N^L_{21} \cos\zeta|^2 + |N^L_{21} \sin\zeta + N^L_{22} \cos\zeta|^2 \\
& + |Q^L_{11} \sin\zeta + Q^L_{21} \cos\zeta|^2 + |Q^L_{21} \sin\zeta + Q^L_{22} \cos\zeta|^2 \Big]
\end{aligned} \tag{A.33}
$$

The squeezing-angle projection weighted C-coefficients are

$$C^{L\theta}_{11} = C^L_{11} \cos\theta + C^L_{12} \sin\theta \tag{A.34}$$
$$C^{L\theta}_{12} = C^L_{12} \cos\theta - C^L_{11} \sin\theta \tag{A.35}$$
$$C^{L\theta}_{21} = C^L_{21} \cos\theta + C^L_{22} \sin\theta \tag{A.36}$$
$$C^{L\theta}_{22} = C^L_{22} \cos\theta - C^L_{21} \sin\theta \tag{A.37}$$

with optimal squeezing angle, Eq. A.33 becomes

$$
\begin{aligned}
S^{\zeta}_{h-DRLSQZ} = \mathcal{Q}^{L} \times \Big[ & e^{-2R}(|C^{L}_{11} \sin\zeta + C^{L}_{21} \cos\zeta|^{2} + |C^{L}_{21} \sin\zeta + C^{L}_{22} \cos\zeta|^{2}) \\
& + |P^{L}_{11} \sin\zeta + P^{L}_{21} \cos\zeta|^{2} + |P^{L}_{21} \sin\zeta + P^{L}_{22} \cos\zeta|^{2} \\
& + |N^{L}_{11} \sin\zeta + N^{L}_{21} \cos\zeta|^{2} + |N^{L}_{21} \sin\zeta + N^{L}_{22} \cos\zeta|^{2} \\
& + |Q^{L}_{11} \sin\zeta + Q^{L}_{21} \cos\zeta|^{2} + |Q^{L}_{21} \sin\zeta + Q^{L}_{22} \cos\zeta|^{2} \Big]
\end{aligned}
\tag{A.38}
$$

Figure A.3b shows squeezing-enhanced strain sensitivity of the Dual-recycled Michelson with optical losses, with injection at various squeezing angles.

## A.3 Parameters Used in Calculations for the Dual-Recycled Michelson

The parameter values used in the quantum noise calculations for the Dual-recycled Michelson are from nominal parameters for the Advanced LIGO interferometer [6–8]. These are listed in Table A.1.

# Appendix B
# Initial Analysis and Repair of the Enhanced LIGO H1 Output Mode Cleaner

**Abstract** This Appendix details the initial diagnostic tests and repairs performed on the Enhanced LIGO H1 Output Mode Cleaner. After a brief introduction to the role and schematic of the Output Mode Cleaner (from [9]), the tests and resulting findings are reported. This initial work was completed by the author, in collaboration and under the supervision of Sam Waldman, and reported in LIGO Technical Note T1100562 [10].

## B.1 DC Readout and the Role of the OMC

The first generation detectors used a RF-readout technique to successfully achieve their design strain-sensitivities [11, 12]. The RF-readout mixes the gravitational-wave signal with RF sidebands to make a heterodyne-type measurement. However, imperfections in the interferometer optics leads to the production of higher-order spatial modes, particularly evident for the RF sidebands used for detection [13]. These higher order spatial modes are detrimental to interferometer performance, contributing additional photon shot noise, increasing the power that needs to be detected and noise couplings, without contributing to gravitational-wave signal.

*DC readout* creates a homodyne-type local oscillator, by putting a small Michelson-length offset to move slightly off the dark fringe point (from Chap. 9—the dark fringe offset). The resulting carrier light at the AS port acts as a Local Oscillator field, noise-filtered by the response of the interferometer coupled-cavities. This is an improvement over using the un-filtered RF sidebands in RF-readout. An installed Output Mode Cleaner filters higher-order spatial modes before the signal is detected. The OMC also removes the RF sidebands, which are still needed to control the interferometer, but would only be detrimental to the DC readout signal. More crucial to the work of this thesis, squeezed vacuum injection is more feasible in conjunction with DC readout than with RF readout, since it requires squeezing in only the audio band rather than at both audio and RF frequencies [14].



S.S.Y. Chua, *Quantum Enhancement of a 4 km Laser Interferometer Gravitational-Wave Detector*, Springer Theses,
DOI 10.1007/978-3-319-17686-4

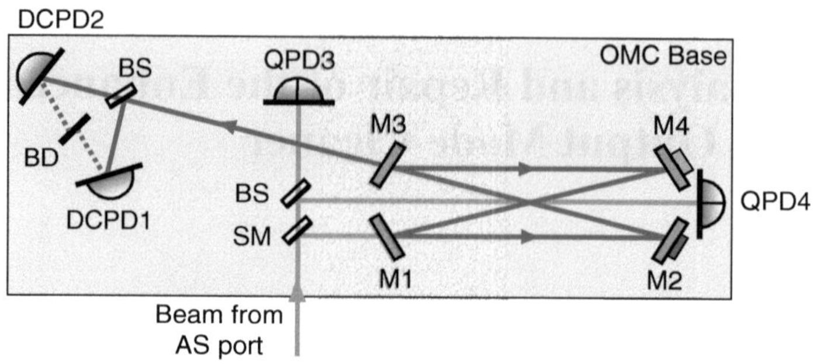


**Fig. B.1** Schematic layout of the OMC glass breadboard. *M1* Flat input coupler, T = 8368 ppm; *M2* PZT mirror with ROC = 2 m, T = 25 ppm; *M3* Flat output coupler, T = 8297 ppm; *M4* OTAS mirror with ROC = 2 m, T = 33 ppm; *SM* Steering mirror, T = 3000 ppm; *BS* 50/50 beamsplitter; *QPD* Quadrant photodetector; *DCPD* DC photodetector, *BD* Beam dump. Adapted from [15]

## B.2 Construction and Design of the OMC

Figure B.1 shows the schematic of the OMC [15]. The Output Mode Cleaner consists of a four-mirror cavity in bow-tie configuration, chosen to prevent direct reflection of rejected light back into the interferometer.

The cavity is constructed by rigidly mounting the cavity optics and photodetectors to a baseplate. The baseplate is a slab of Corning ULE glass, with the components bonded using UV-cure epoxy. Two of the cavity mirrors have cavity-length actuators: a fast, short-range ($<0.1\,\mu$m) PZT, and a slow, long-range ($\sim 20\,\mu$m) thermal actuator, consisting of a 1 in. long segment of aluminum tube warmed by a resistive heater. The slow actuator is named the *OMC Thermal Actuator System* (*OTAS*).

To isolate the output mode cleaner from environmental disturbances, the glass slab bench was hung from an actively-damped double-pendulum suspension system [16, 17], which was in turn was mounted on an in-vacuum isolation system.

## B.3 Transmission Loss of the H1 OMC and the Effect on Realised Squeezing Magnitude

The transmission of the H1 OMC dropped suddenly during the joint science run [18], by approximately 25 % lower transmission. The power drop was correlated with a electric power outage at the Hanford Observatory, and finally connected with a decrease in the OMC Finesse, from the original $\mathscr{F} = 367$ to $\mathscr{F} = 297$ [19]. In this condition, the OMC only transmitted 66 % of the input light. This became a limiting optical loss source, a problem for the LIGO Squeezed Light Injection Experiment, limiting the observable squeezing to 1dB [20].

### *B.3.1 LIGO Squeezed Light Injection Experiment—OMC Swap-Out*

With the OMC from LIGO Livingston Observatory interferometer (L1) at hand in Hanford, it was undertaken to swap out the H1 OMC with the L1 OMC. After measurement of the L1 OMC to have a Finesse of 369.8 [21] in a bench-top experiment, the LIGO Squeezed Light Injection team replaced the H1 OMC with the L1 OMC [22]. After the swap-out, the H1 OMC was tested in a bench-top experiment to determine the cause of the additional loss.

## B.4 Experiment Setup

Two sets of data are needed to diagnose the OMC. Firstly, the OMC transmission and mode matching is required. Secondly, the OMC finesse is needed. The transmission measurements were performed with a laser locked to the OMC, while the Finesse measurements were made by sweeping over cavity resonances via two techniques.

### *B.4.1 Test Setup*

The experiment schematic is shown in Fig. B.2. The laser (Lightwave 126 NPRO) was locked to the OMC using a PDH locking loop, feeding back into the "Fast" input actuator. The phase modulation was generated by a resonant EOM (New Focus), driven by a RF amplifier. The modulation frequency was 21.5 MHz. The OMC reflected and transmitted signals were measured with photodetectors (Thorlabs PDA10CS). The reflected signal was demodulated and low pass filtered. The low-passed signal was then boosted and further filtered by a pre-amplifier unit (Stanford Research Systems SR560), before feeding into the servo controller (New Focus LB1005). The error signal had a range of $\approx\pm 10\,V_{pp}$. To match the 0–10 V Thorlabs piezo controller unit input, the servo controller output was summed with a 9 V battery and a 10 kΩ resistor, housed in a electronics box (Pomona). The resistor, in combination with the 10 kΩ input impedance of the piezo controller unit, converted the $\pm 10$ V servo controller output to the 0–10 V piezo controller input signal range.

### *B.4.2 Transmission Measurements*

The transmission measurements were made with the laser locked to the cavity. Photodetectors were installed at the input (via a pickoff mirror), in reflection and in

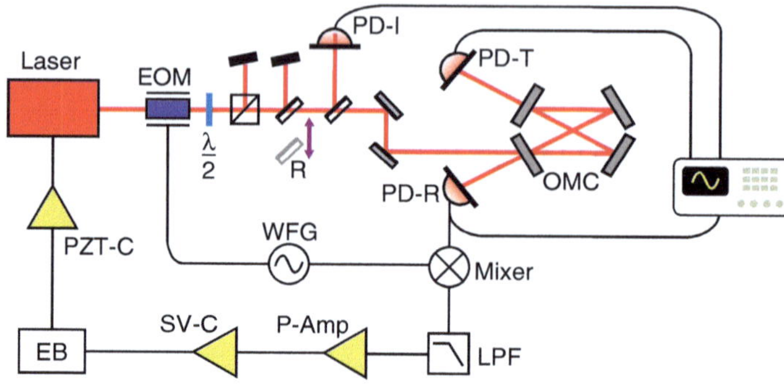


**Fig. B.2** Experiment configuration for the H1 OMC test. *Laser* NPRO laser; *EOM* Electro-optic modulator; $R = 98\,\%$ mirror (for low power measurements); *PD-(I, R, T)* Photodetector-(Input, Reflection, Transmission); *WFG* Waveform generator; *LPF* Low-pass filter; *P-Amp* Pre-amplifier unit; *SV-C* Servo controller unit; *EB* Electronics box; *PZT-C* Piezo controller unit

transmission of the OMC. All three photodiode readings were recorded while the OMC was both on and off resonance, to measure the power transmission.

### *B.4.3 Finesse Measurements*

The finesse was measured via two methods, named (for this work) *slow* and *fast* methods. In the slow method, the laser temperature was stepped by a large amount so that the OMC would sweep through multiple resonances. The resonance signal was recorded and the fringes fitted offline using a Matlab program. In the fast method, the laser PZT was swept at 1, 10 or 100 Hz with a ramp waveform across a single resonance and PDH sidebands. The time between the carrier and RF sidebands was read off directly from an oscilloscope, as was the FWHM of the carrier resonance. Together, these allow for a calculation of the OMC finesse using the previously measured FSR of 278 MHz [15].

The slow method had a systematic error source. Digitisation noise of the oscilloscope limited the resolution of the voltage measurements that were taken. This gave skewed finesse values—mostly low readings. Instead, the spacing between the carrier and RF sidebands, as well as the width of the FWHM, must be measured as a *time* measurement, using the fast method as described above.

### *B.4.4 Power Scaling*

The power incident on the OMC was controlled using a waveplate and a polarising-beam-splitter (PBS) cube. For small power levels, a 98 % beamsplitter was used to divert most of the light.

## B.5 Measurement Results

### *B.5.1 Transmission/Throughput*

The data taken for the transmission measurement is summarised in Table B.1. Thus the power transmitted through the cavity is 0.36/0.38 = 95 %.

### *B.5.2 Cavity Finesse*

The finesse was measured with 0.5 mW incident on the OMC. The Finesse was measured using the 'fast' method with a ramp of 1 and 10 Hz, and comparable values found for both measurements. For the 10 Hz sweep, the time between the sideband resonances was 17.68 ms, while the FWHM of the resonance was 336 μs. Using the sidebands as a frequency calibration, the FWHM is thus $0.336/17.7 \times 2 \times 21.5\,\text{MHz} = 0.82\,\text{MHz}$. With a free spectral range of 278 MHz, the finesse was determined to be $\mathscr{F} = 278/0.82 = 340$.

### *B.5.3 Power scaling*

The OMC transmission was measured as a function of the input power, presented in Fig. B.3. This shows the transmission having a clear dependence on the input power. The measurement setup was adjusted to ensure no other experimental effects were the cause (such as changes in alignment or modematching), and measurements repeated multiple times across different optical input powers. However, the overall result remained consistent.

**Table B.1** Transmission/throughput data

| Parameter | Value (mW) |
|---|---|
| Input power | 0.55 |
| Transmitted power | 0.36 |
| Reflected power (locked) | 0.16 |
| Reflected power (unlocked) | 0.54 |

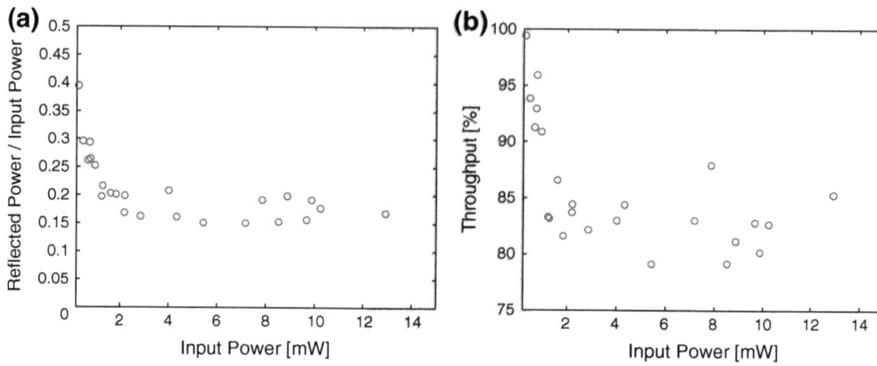


**Fig. B.3** OMC power scaling experiment data. **a** Normalised reflected power as a function of incident input power, indicating the input beam is ~83 % spatially modematched to the OMC. **b** Throughput measurement (comparing measured power entering and exiting the OMC) as a function of the incident input power. The OMC throughput/transmission depends strongly on the power incident on and circulating in the cavity

## B.6 Discussion and Further Work

The finesse and transmission are related by

$$\frac{P_T}{P_0} = \left(\frac{T\mathscr{F}}{\pi}\right)^2 \tag{B.1}$$

The H1 OMC had a coupler transmissivity of $T = 8300$ ppm [15]. Using Eq. B.1

- For the Finesse of 340, the calculated transmission is only $P_T/P_0 = 80\,\%$.
- Alternatively, the transmission of 95 % implies a Finesse of 370.

Although these numbers are in contradiction, it is likely that the FWHM reported above is an overestimate due to motion of the laser frequency during the measurement sweep, and the Finesse is, in practice, closer to 370 than to 340.

The loss of a cavity is calculated from the finesse and transmission by

$$l = 1 - \left(\frac{1}{1-T}\left(1 - \frac{\pi}{\mathscr{F}}\right)^2\right) \tag{B.2}$$

As shown in Fig. B.4, 95 % transmission corresponds to a round trip loss of 400 ppm, while 84 % transmission corresponds to 1400 ppm. The power dependence of the loss is possibly due to beam clipping changing with temperature. One source candidate of such a loss (both the static and power dependent), is beam clipping at the thermal actuator (OTAS). To account for such an increase in loss via this mechanism, this would require:

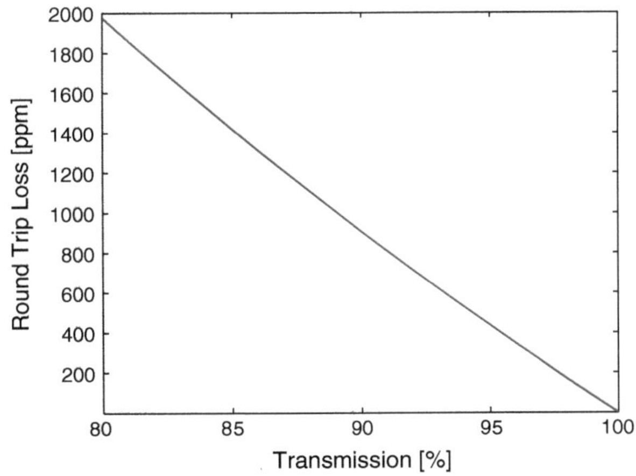


**Fig. B.4** The round trip loss of the OMC as a function of the transmission

- Translating a hard edged aperture by almost 100 μm. With the OTAS tube made of aluminum, the coefficient of thermal expansion of Aluminium is 23 ppm/K [23]. Assuming a 1 mm zone of influence, the temperature rise required to cause such a hard-edge translation is an unfeasible and unphysical 4600 K.
- A change in cavity-mode mismatching of 10 %. The cavity mode would require a change in the waist size from ≈150 μm with respect to the nominal ≈500 μm waist size. This would require a change in the PZT and OTAS mirror radii-of-curvature from R = 2 m to R = 1 m, also unfeasible and unphysical.

The origin of the increased loss in the H1 OMC at high incident optical power remains an open question, however, the initial analysis and test setup reported here have confirmed the findings made during the joint science run. The next stage is to reconstruct the testing rig used to initially characterise the OMC. This testing setup, detailed in Sect. 3.3 of [24], will allow Finesse and transmission measurements with greater precision via using a well-controlled acousto-optic modulator to sweep input beam frequencies.

# Appendix C
# Supplementary Information for the Backscatter-Light Experiments at LIGO

**Abstract** This Appendix provides supplementary information for Chap. 10—*Back-scattered-light impact in a squeezing-enhanced gravitational-wave detector*. Details of intermediate hardware improvements and additions to the squeezer apparatus that reduced backscattered-light coupling are covered. The small-displacement measurement results made before the intermediate improvements are then presented. The demonstration of the driven RIN equation used in the large displacement measurement as a spectrum, and calculation for the modulation depth conclude this Appendix.

## C.1 Hardware Improvements to Reduce Backscattered-Light Coupling

Intermediate improvements and additions were made to the LIGO squeezer experimental setup during the Experiment window. They aimed to reduce both the vibrational motion of squeezing injection path optics, as well as the (DC) optical power of the backscattered beam, for backscattered light mitigation.

### *C.1.1 Optics Motion Reduction*

Squeezing injection path optics were in-situ tested for resonances. This was achieved by observing the interferometer readout spectrum in real time while the optical mounts were bowed to individually excite angular and flagpole-type motions. Of the twenty individual optics in the injection path investigated, only the 2″ steering mirror mounts produced peaks in the readout. The other optics, including the transmissive elements (lenses, half-wave plate) did not produce peaks.

Assuming flagpole resonances were the dominant motion, clean foam pads weighted with metal bolts were placed on top of and supported by multiple optics [25]. This damping of optics is shown Fig. C.1a.

Shown in Fig. C.1b, measurements of the interferometer readout with applied damping foam (blue traces) and without (red traces) were made in repeat-cycle.



S.S.Y. Chua, *Quantum Enhancement of a 4 km Laser Interferometer Gravitational-Wave Detector*, Springer Theses,
DOI 10.1007/978-3-319-17686-4

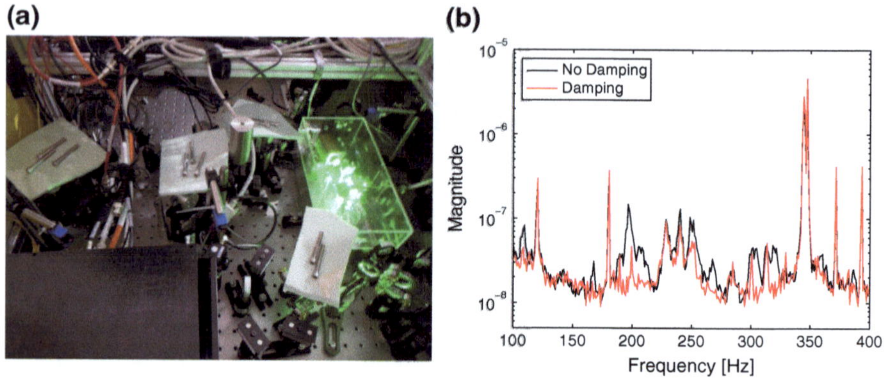


**Fig. C.1** **a** Weighted damping material placed upon mirror mounts in the squeezing injection path. **b** Improvement in the interferometer readout spectrum with damping (*red*) versus without damping (*black*)

Peak reduction was seen, particularly evident at 195, 203, 267 and 320 Hz. Peaks at 230, 240, and 250 Hz, caused by steering mirrors undamped during the repeat-cycle measurement, were subsequently damped in a similar fashion.

## *C.1.2 Scattered Beam Power Reduction*

The presence of scatter-beam retro-reflections retain DC light power that would be otherwise lost to the propagation mode. Sources of retro-reflections include mirror mount back surfaces, silvered reverse-side mirror backs, and coincidental normal incidence on reflective or scattering surfaces.

To search for possible retro-reflection, an Optical Parametric Oscillator with no pump beam and no auxiliary beam incident was cavity-length scanned while observing the interferometer readout. This caused peaks at the scanning frequency to be generated in the readout spectrum. A black glass plate was then used to probe the squeezed-light input path in reverse-injection direction. This plate was systematically placed after each optical element.

The main retro-reflection found occurred where scattered light coming from the interferometer up the injection chain was reflecting off the OPO, then retro-reflecting on an on-table IR Camera via the dichroic beamsplitter in the OPO pump injection path [26]. This is illustrated in Fig. C.2a.

To block this retro-reflection, a small black glass 'V' beam dump was placed in front of the IR Camera. The reduction in the generated peak heights in the readout spectrum with the blocked retro-reflection is shown in Fig. C.2b. All measurement conditions were kept constant, while the beam dump was placed into and removed from the retro-reflection beam path. This process was cycled three times with similar

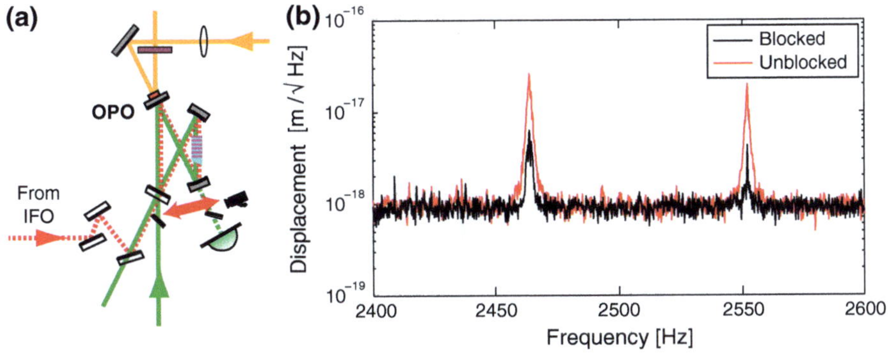


**Fig. C.2** **a** Simplified schematic of the squeezing injection beam path. The scattered-light retro-reflection is shown by the *thick double arrow*. A small *black glass* beam dump was used to block the retro-reflection. **b** Generated peaks in the readout spectrum with retro-reflection unblocked (*red*), and reduced peak heights with retro-reflection blocked (*black*)

reduction results, thus confirming the real effect of removing DC power from the backscattered-light beam.

## C.1.3 Reduced Backscattered-Light Noise Coupling to Readout

A comparison of readout spectrum and correlation measurements made before and after the improvements is shown in Fig. C.3. The readout spectra (with injected squeezing blue traces versus no-squeezing reference red traces) after the improvements shows a reduction of structures that introduce noise over the reference trace, particularly between the 200 to 300 Hz frequency region.

Noise correlation data was measured between the interferometer readout signal and each of the on-table accelerometers (X, Y, Z) signals, using environmental vibrations. An approximate reduction in coherence between 200 and 300 Hz of 25 %, confirms the reduction of structures is a real effect.

These improvements and additions were made during the LIGO Squeezed Light Injection Experiment Window. One backscatter measurement were made prior to these improvements—a small displacement measurement with both shaker and acoustic injections. Due to time constraints in the Experiment Window scheduling, the acoustic small displacement measurement were not repeated after these improvements were made. The result is presented next, but due to the differing Experiment conditions, they cannot be compared to the final backscatter measurements presented in Chap. 10.

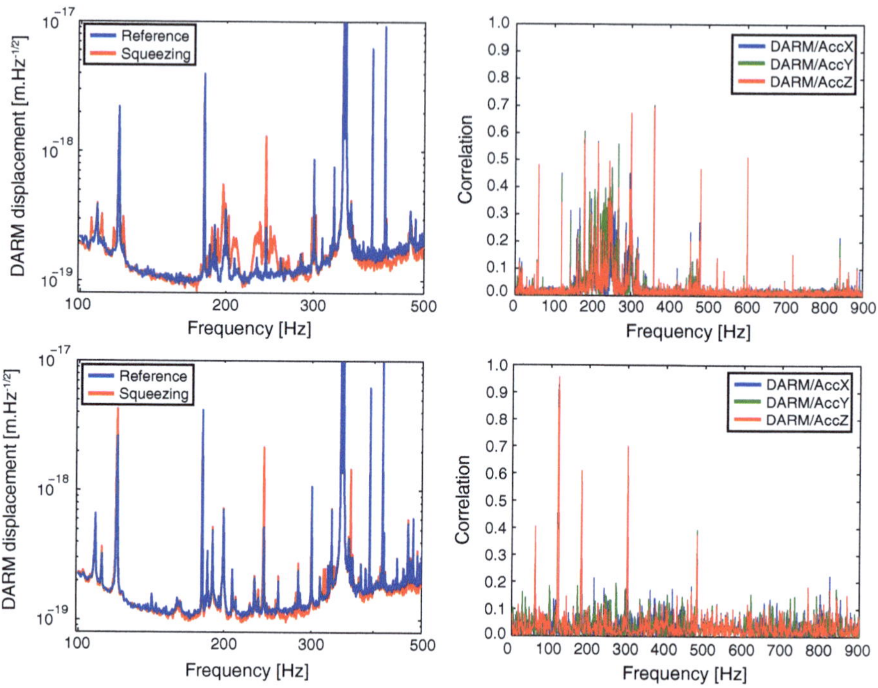


**Fig. C.3** *Left column* Interferometer readout spectra before and after the improvements were made, with feature reduction evident in the 200–300 Hz region. *Right column* Correlation between the readout signal and on-table accelerometers signals, taken with ambient vibration only

## C.2 Small Displacement Measurement with Shaker and Acoustic Injections

This small displacement measurement has the same concepts behind the measurement reported in Chap. 10, but included different types of shaker injections, as well as an Acoustic injection [27]. The inferred backscatter noise level for this test is shown in Fig. C.4. The various data points are as follows:

- **Acoustic 10 Hz comb injection:** This was accomplished by broadcasting a 10 Hz comb signal using a loud speaker near the squeezer table. The displacement readout was a microphone placed near the squeezer table. The inferred noise levels are shown by the red-circle points above 100 Hz (not every 10 Hz comb point is displayed, for clarity).
- **Piezo shaker 10 Hz comb injection:** A 10 Hz frequency comb signal was injected by the piezo shaker on the squeezer table, with the motion readout by the accelerometers. The inferred noise levels for these injections are shown by the square-points below 100 Hz.

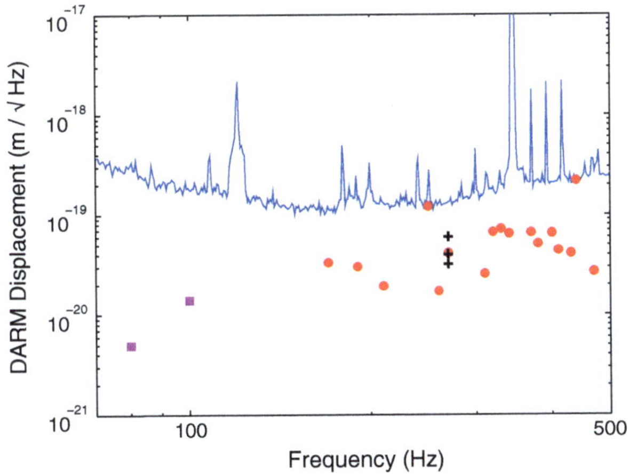


**Fig. C.4** Small-displacement measurement prior to intermediate backscatter improvements. This data is presented in units of interferometer readout displacement, not normalised RIN. *Circles* Acoustic 10 Hz comb injection, above 100 Hz (not every 10 Hz comb point has been displayed for clarity); *Squares* Piezo shaker 10 Hz comb injection, below 100 Hz; *Crosses* Single frequency PZT shaker injections at 270 Hz. The three data points were made with three different shaking amplitudes varying by a factor of about 10.

- **Single frequency PZT shaker injections:** These are the three black-cross points at 270 Hz frequency. The three data points were made with three different shaking amplitudes varying by an order of magnitude.

Two important features are shown by Fig. C.4. Firstly, prior to the intermediate hardware improvements, certain inferred backscatter noise level points were comparable to the interferometer readout spectrum. This characteristic is consistent with the coherence measurements showing correlated features. Secondly, the single-frequency PZT shaker injections (black-cross points at 270 Hz) result in inferred values that vary by a factor of 2, despite the change in driving amplitude to vary by an order of magnitude. This verifies, to within measurement error, that small displacements scale linearly.

## C.3 The Large Displacement Measurement Equation—A 'Shelf' Spectrum

This section demonstrates the large displacement measurement equation is a 'shelf-spectrum'. This derivation follows the work presented by Fricke [28]. We first define the angular drive frequency from the linear drive frequency $f_{\text{drive}}$:

$$\Omega_{\text{dr}} = 2\pi f_{\text{dr}} \tag{C.1}$$

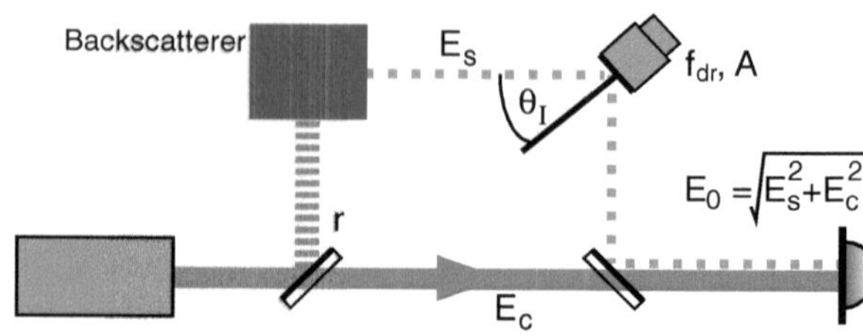


**Fig. C.5** Block diagram/schematic for the large displacement measurement

We also define the modulation depth $M$:

$$M = \frac{2\pi}{\lambda}\frac{2A}{\cos\theta_{\mathrm{I}}} \tag{C.2}$$

where $A$ is the applied motion in metres and $\theta_{\mathrm{I}}$ is the angle of incidence onto the driven optic.

The electric field of the backscattered beam $E_s$ and the un-scattered field $E_c$ can be described by:

$$E_s = rE_0 \exp\{i\mathrm{M}\sin(\Omega_{\mathrm{dr}}\mathrm{t})\} \tag{C.3}$$

$$E_c = (1-r)E_0 \tag{C.4}$$

where $E_0$ is the total field incident onto the photodetector, and $r$ is the proportion of $E_0$ in the backscattered field. Figure C.5 shows the block diagram/schematic and the different variables. The photodetector detects the power of incident field, thus:

$$\begin{aligned} P &= |E_s + E_c|^2 \\ &= |E_s|^2 + |E_c|^2 + 2\mathbf{Re}E_sE_c^* \\ &= P_0 + 2\mathbf{Re}E_sE_c^* \end{aligned} \tag{C.5}$$

Substituting Eqs. C.3 and C.4 into Eq. C.5 gives:

$$\begin{aligned} P &= P_0 + 2[(1-r)E_0^*]rE_0\mathbf{Re}\exp\{i\mathrm{M}\sin(\Omega_{\mathrm{dr}}\mathrm{t})\} \\ &= P_0 + 2(1-r)rP_0\mathbf{Re}\exp\{i\mathrm{M}\sin(\Omega_{\mathrm{dr}}\mathrm{t})\} \end{aligned} \tag{C.6}$$

We now substitute the Bessel function expansion of $\exp\{i\mathrm{M}\sin(\Omega_{\mathrm{dr}}\mathrm{t})\}$, thus:

$$P(f) = P_0 + 2r(1-r)P_0\mathbf{Re}\sum_{n=-\infty}^{\infty} J_n(\mathrm{M})e^{\mathrm{in}\Omega_{\mathrm{drive}}\mathrm{t}} \tag{C.7}$$

Utilising the Bessel function symmetry property $J_{-n}(z) = (-1)^n J_n(z)$ (Eq. 6, Sect. 2.11 of [29]), we can rearrange the summation term to the following:

$$\sum_{n=-\infty}^{\infty} J_n(\mathrm{M})e^{in\Omega_{\mathrm{dr}}t} = J_0(\mathrm{M}) + \sum_{n=-\infty}^{-1} J_n(\mathrm{M})e^{in\Omega_{\mathrm{dr}}t} + \sum_{n=1}^{\infty} J_n(\mathrm{M})e^{in\Omega_{\mathrm{dr}}t} \quad (\mathrm{C.8})$$

$$= J_0(\mathrm{M}) + \sum_{n=1}^{\infty} \underbrace{J_n(\mathrm{M})\left((-1)^{\mathrm{n}}\mathrm{e}^{-\mathrm{i}n\Omega_{\mathrm{dr}}\mathrm{t}} + \mathrm{e}^{\mathrm{i}n\Omega_{\mathrm{dr}}\mathrm{t}}\right)}_{Q} \quad (\mathrm{C.9})$$

This is a *power spectrum*. The spectrum is made of a comb of peaks at multiples of $\Omega_{\mathrm{dr}}$, with the peak amplitudes given by Bessel functions $J_n(\mathrm{M})$ where $n = f/f_{\mathrm{dr}}$. Any noise on the field will get superimposed on each of the peaks, smoothing out the spectrum, creating a plateau.

The 'shelf' cutoff occurs because the amplitude of Bessel functions $J_n(\mathrm{M})$ rolls off steeply with $n$ when $n > M$. This creates the 'shelf' structure used in the Large displacement measurement in Chap. 10. The cutoff frequency, or *knee frequency* $f_{\mathrm{knee}}$ is where $n \approx M$. Thus, the knee frequency will be

$$N = \frac{f_{\mathrm{knee}}}{f_{\mathrm{dr}}} = M$$
$$\Rightarrow f_{\mathrm{knee}} = Mf_{\mathrm{dr}} \quad (\mathrm{C.10})$$

which corresponds to Eq. 10.10. Returning to Eq. C.9, focussing on the $Q$ term, for odd integers of $n$:

$$J_n(\mathrm{M})(e^{in\Omega_{\mathrm{dr}}\mathrm{t}} - e^{-in\Omega_{\mathrm{dr}}\mathrm{t}}) = i \times 2J_n(\mathrm{M})\sin(n\Omega_{\mathrm{drive}}t) \quad (\mathrm{C.11})$$

while for even integers of $n$:

$$J_n(\mathrm{M})(e^{in\Omega_{\mathrm{drive}}t} + e^{-in\Omega_{\mathrm{drive}}t}) = 2J_n(\mathrm{M})\cos(n\Omega_{\mathrm{dr}}t) \quad (\mathrm{C.12})$$

Hence, evaluating the real component gives:

$$\mathbf{Re}\exp\{i\mathrm{M}\sin(\Omega_{\mathrm{drive}}t)\} = J_0(\mathrm{M}) + 2\sum_{n=1}^{\infty} J_{2n}(\mathrm{M})\cos(2n\Omega_{\mathrm{dr}}t) \quad (\mathrm{C.13})$$

Now, using Jacobi-Anger Identity (Eq. 4, Sect. 3.11 of [29]):

$$\cos(x\sin y) = J_0(x) + 2\sum_{n=1}^{\infty} J_{2n}(x)\cos(2ny) \quad (\mathrm{C.14})$$

where $x \to M$ and $y \to \Omega_{\mathrm{dr}}t$, and further simplifying by noting:

$$\lim_{r\to 0}(1-r)r = r \quad (\mathrm{C.15})$$

and

$$r = \frac{E_s}{E_0} \approx \frac{E_s}{E_c} \tag{C.16}$$

The power spectrum expression of Eq. C.6 now becomes:

$$P(f) \approx P_0 + 2\frac{E_s}{E_c}P_0 \cos(\mathrm{M} \sin \Omega_{\mathrm{dr}}\mathrm{t}) \tag{C.17}$$

Thus

$$P(f) = P_0 + 2\sqrt{\frac{P_s}{P_c}}\, P_0 \cos(\mathrm{M} \sin \Omega_{\mathrm{dr}}\mathrm{t}) \tag{C.18}$$

It then follows that the relative intensity noise (RIN) is given by:

$$RIN = \frac{\mathrm{d}P(f)}{P_0} = 2\sqrt{\frac{P_n}{P_c}} \cos(\mathrm{M} \sin \Omega_{\mathrm{dr}} t) \tag{C.19}$$

which is the Large displacement measurement Equation (Chap. 10, Eq. 10.4(i)). As a final note, one can bypass the Bessel function expansion and Jacobi-Anger Identity by using Euler's Formula:

$$\begin{aligned} \exp\{i\mathrm{M} \sin(\Omega_{\mathrm{dr}} t)\} &= \cos(\mathrm{M} \sin(\Omega_{\mathrm{dr}} t)) + i \sin(\mathrm{M} \sin(\Omega_{\mathrm{dr}} t)) \\ \mathbf{Re} \exp\{i\mathrm{M} \sin(\Omega_{\mathrm{dr}} t)\} &= \cos(\mathrm{M} \sin \Omega_{\mathrm{dr}} t) \end{aligned} \tag{C.20}$$

thus going from Eqs. C.6 to C.17, however, the Bessel function expansion methodology explicitly demonstrates that it is indeed a power *spectrum*.

## C.4 Calculation of the Modulation Depth for the Large Displacement Measurement

The calculation of the modulation depth ($M$) from known parameters is summarised in Table C.1.

**Table C.1** Summary of modulation depth calculation

| Name | Variable | Value | Derivation |
|---|---|---|---|
| PZT response | $Spec_{\mathrm{PZT}}$ | 30 μm/10,00 V | Product specifications |
| TREK output/input drive | $Spec_{\mathrm{TREK}}$ | 100 V/V | Product specifications |
| PZT drive | $Drive_{\mathrm{PZT}}$ | 3 μm/V | $Spec_{\mathrm{PZT}} \times Spec_{\mathrm{TREK}}$ |
| Path length change/input drive | $\Delta L/V$ | 3.1 μm/V | $Drive_{\mathrm{PZT}} \times (\cos(50))^{-1}$ |
| Applied drive voltage | $V_{\mathrm{applied}}$ | 3 $\mathrm{V_{pp}}$ | Experiment value |
| Modulation depth | $M$ | 173 rad | $V_{\mathrm{applied}} \times \Delta L/V \times 4\pi/\lambda$ |

# References


1. A. Buonanno, Y. Chen, Quantum noise in second generation, signal-recycled laser interferometric gravitational-wave detectors. Phys. Rev. D **64**, 042006 (2001)
2. A. Buonanno, Y. Chen, Improving the sensitivity to gravitational-wave sources by modifying the input-output optics of advanced interferometers. Phys. Rev. D **69**, 102004 (2004)
3. V. Chickarmane, S.V. Dhurandhar, T.C. Ralph, M. Gray, H.-A. Bachor, D.E. McClelland, Squeezed light in a frontal-phase-modulated signal-recycled interferometer. Phys. Rev. A **57**, 3898–3912 (1998)
4. J. Harms, Y. Chen, S. Chelkowski, A. Franzen, H. Vahlbruch, K. Danzmann, R. Schnabel, Squeezed-input, optical-spring, signal-recycled gravitational-wave detectors. Phys. Rev. D **68**, 042001 (2003)
5. Gravitational Wave Interferometer Noise Calculator (GWINC) v3, https://awiki.ligo-wa.caltech.edu/aLIGO/GWINC
6. R. Abbott, R. Adhikari, S. Ballmer, L. Barsotti, M. Evans, P. Fritschel, V. Frolov, G. Mueller, B. Slagmolen, S. Waldman, Advanced LIGO length sensing and control final design (2010). LIGO-T1000298-T
7. M.A. Arain, C. Mueller, G. Mueller, Optical layout and parameters for the advanced LIGO cavities (2009). LIGO-T0900043-11
8. P. Fritschel, R. Adhikari, S. Ballmer, M. Evans, Arm cavity finesse for advanced LIGO (2007). LIGO-T070303-01-D
9. T.T. Fricke, N.D. Smith-Lefebvre, R. Abbott, R. Adhikari, K.L. Dooley, M. Evans, P. Fritschel, V.V. Frolov, K. Kawabe, J.S. Kissel, B.J.J. Slagmolen, S.J. Waldman, DC readout experiment in Enhanced LIGO. Class. Quantum Gravity **29**(6), 065005 (2012)
10. S. Chua, S. Waldman, eLIGO hanford output mode cleaner repair (2011). LIGO-T1100562-v2
11. The LIGO Scientific Collaboration, (B. P. Abbott et. al.), LIGO: the laser interferometer gravitational-wave observatory. Rep. Prog. Phys. **72**, 076901 (2009)
12. T. Accadia et al., Status of the Virgo project. Class. Quantum Gravity **28**, 114002 (2011)
13. A.M. Gretarsson, E. D'Ambrosio, V. Frolov, B. O'Reilly, P.K. Fritschel, Effects of mode degeneracy in the LIGO Livingston Observatory recycling cavity. J. Opt. Soc. Am. B **24**(11), 2821–2828 (2007)
14. J. Gea-Banacloche, G. Leuchs, Squeezed states for interferometric gravitational-wave detectors. J. Mod. Opt. **34**(6–7), 793–811 (1987)
15. S. Waldman, The Enhanced LIGO output mode cleaners (2009). LIGO-T080144-v1
16. N.A. Robertson, D. Coyne, C. Echols, P. Fritschel, V. Mandic, J. Romie, C. Torrie, S. Waldman, Conceptual design of a double pendulum for the output modecleaner (2006). LIGO-T060257-03-R
17. N.A. Robertson, M. Barton, D. Bridges, J. Heefner, M. Meyer, J. Romie, C. Torrie, OMC suspension final design document (2009). LIGO-T0900060-v1
18. N. Smith, OMC transmission is low again. Hanford iLog 8, (Sept 2010)
19. K. Kawabe, OMC finesse and transmission. Hanford iLog 5, (Aug 2011)
20. L. Barsotti, Too many losses: data with H1 OMC. Hanford iLog 26, (Oct 2011)
21. S. Chua, L1 OMC inspection and resonance scan. Hanford iLog 21, (Oct 2011)
22. N. Smith, OMC resonates. Hanford iLog 25, (Oct 2011)
23. A.J.C. Wilson, The thermal expansion of aluminium from 0 °C to 650 °C. Proc. Phys. Soc. **53**(3), 235 (1941)
24. N. Smith-Lefebvre, Techniques for improving the readout sensitivity of gravitational wave antennae. Ph.D. thesis, Department of Physics, Massachusetts Institute of Technology, Cambridge, 2012

25. R. Schofield, Damping of DARM peaks from squeezing. Hanford iLog 14, (Nov 2011)
26. S. Chua, Stray scattered light hunting. Hanford iLog 19, (Nov 2011)
27. R. Schofield, The impact of squeezing on vibration coupling to H1. Hanford iLog 23, (Oct 2011)
28. T. Fricke, Homodyne detection for interferometric gravitational-wave detectors. Ph.D. thesis, Graduate Faculty, Louisiana State University, 2011
29. N.W. McLachlan, *Bessel Functions for Engineers* (Oxford University Press, Oxford, 1961)

# Index




S.S.Y. Chua, *Quantum Enhancement of a 4 km Laser Interferometer Gravitational-Wave Detector*, Springer Theses,
DOI 10.1007/978-3-319-17686-4

MIX
Papier aus verantwortungsvollen Quellen
Paper from responsible sources
FSC® C105338

If you have any concerns about our products,
you can contact us on
**ProductSafety@springernature.com**

In case Publisher is established outside the EU,
the EU authorized representative is:
**Springer Nature Customer Service Center GmbH**
**Europaplatz 3, 69115 Heidelberg, Germany**

Printed by Libri Plureos GmbH
in Hamburg, Germany